INVESTIGATING LIFE CYCLE CARBON EMISSION IMPACT OF ROAD INVESTMENTS

OCTOBER 2025

ASIAN DEVELOPMENT BANK

ADB

6 ADB Avenue, Mandaluyong City, 1550 Metro Manila, Philippines
Tel +63 2 8632 4444; Fax +63 2 8636 2444
www.adb.org

ISBN 978-92-9277-420-2 (print); 978-92-9277-421-9 (PDF); 978-92-9277-422-6 (ebook)
Publication Stock No. TCS250338-2
DOI: http://dx.doi.org/10.22617/TCS250338-2

Notes:
In this publication, "$" refers to United States dollars.
ADB recognizes "China" as the People's Republic of China and "Kyrgyzstan" as the Kyrgyz Republic.

On the cover: **Vehicles on the road in Georgia.** Vehicles passing through Zviad Gamsakhurdia (Right Bank) Street in Tbilisi (photo by ADB).

CONTENTS

TABLES AND FIGURES

TABLES

FIGURES

FOREWORD

Globally, transport is and will continue to be dominated by road-based modes. Within Asia and the Pacific, substantial road investment is needed to address the infrastructure gap and respond to a growing demand for mobility and increasing vehicle ownership. This provides the opportunity to improve and expand road networks, and to address negative externalities by ensuring greener, safer, climate-resilient, and adaptive designs.

Sector emissions, both local and global, are projected to substantially increase over the coming years. Stringent efforts should be made to reduce emissions and support new pathways to lower and to ultimately decarbonize the transport sector and clean the air.

It is important to recognize that most greenhouse gas emissions from the transport sector come from the use of transport infrastructure, rather than the provision of the infrastructure itself. There are emissions related to building or maintaining transport assets, but these typically account for a small proportion (often well below 10%) of total life cycle-related emissions. It is not the roads that have climate change emissions, but the vehicles that use these roads. As such, activities to address the transport sector's climate change emissions must take a broad view and aim to address the sector activities that emit the most.

Decarbonization of the transport system has started and will eventually lead to a net-zero sector in terms of climate change emissions. The roads of the future will carry zero emission vehicles. What is important is to accelerate the activities to reduce emissions and bend the curve downward in terms of emission growth.

This publication contributes to the above efforts by presenting a comprehensive methodology for life cycle assessment of road transport emissions, accompanied by a tool kit that works with the widely used Highway Development and Management system (HDM-4). The tool can be applied by practitioners to guide decisions on effective avoid–shift–improve measures to reduce emissions and can be used to support the planning and design of road projects and programs that align with the goals of the Paris Agreement on climate change and the 2030 Sustainable Development Agenda.

James Leather
Director
Transport Sector Office
Asian Development Bank

ACKNOWLEDGMENTS

This publication was produced with funding support from TA 6756: Improving Infrastructure Sustainability through Better Asset Management, and the Knowledge Advisory Services Center of the Asian Development Bank (ADB). The study was conceptualized and led by Michael Anyala, senior transport specialist (Road Asset Management). Technical work was carried out by Ian Greenwood (ADB consultant from New Zealand) and Eric Stannard (University of Birmingham, United Kingdom), with administrative support provided by Diana Marie Hernandez-Louis and Franzella Pinky Villanueva of ADB. We gratefully acknowledge the contributions of Frank van Steenbergen and Anastasia Deligianni of MetaMeta, who provided tree planting sequestration rates for the study. Special thanks are extended to the many peer reviewers from across ADB.

ABBREVIATIONS

ADB	Asian Development Bank
CO_2e	carbon dioxide equivalent
PRC	People's Republic of China
DMC	developing member country
EV	electric vehicle
GHG	greenhouse gas
grid factor	CO_2e emissions produced by generating one kilowatt-hour of energy
GWP	global warming potentials
HDM-4	Highway Development and Management system
ICE	internal combustion engine
IPCC	Intergovernmental Panel on Climate Change
LZEV	low- or zero-emission vehicle
FSM	Federated States of Micronesia
N_2O	nitrous oxide
NMT	nonmotorized transport
NPV	net present value
PHEV	plug-in hybrid electric vehicle
PNG	Papua New Guinea
SDG	Sustainable Development Goal
SUV	sports utility vehicle
TTW	tank-to-wheel
UNFCCC	United Nations Framework Convention on Climate Change
WTT	well-to-tank
WTW	well-to- wheel

WEIGHTS AND MEASURES

GJ	gigajoule
kg	kilogram
km	kilometer
m^2	square meter
m^3	cubic meter
$mtCO_2e$	metric tons of carbon dioxide equivalent
tCO_2e	tons of carbon dioxide equivalent

INTRODUCTION 1

Context

The Asian Development Bank (ADB) has committed through Strategy 2030 to work with its developing member countries (DMCs) in tackling climate change, building climate and disaster resilience, and enhancing environmental sustainability through its operations. Linked to this, ADB aims to align its operations with the mitigation and adaptation goals of the Paris Agreement on climate change by ensuring all projects follow a low-carbon and climate-resilient pathway.

Emissions in transport sector is growing faster than those from any other major economic sectors (SLOCAT 2021). Across world regions, growth in transport sector emissions is highest in Asia (Figure 1), driven largely by growing demand in road passenger and freight transport.

Figure 1: Increase in Transport Sector Emissions by Region, 2010–2019

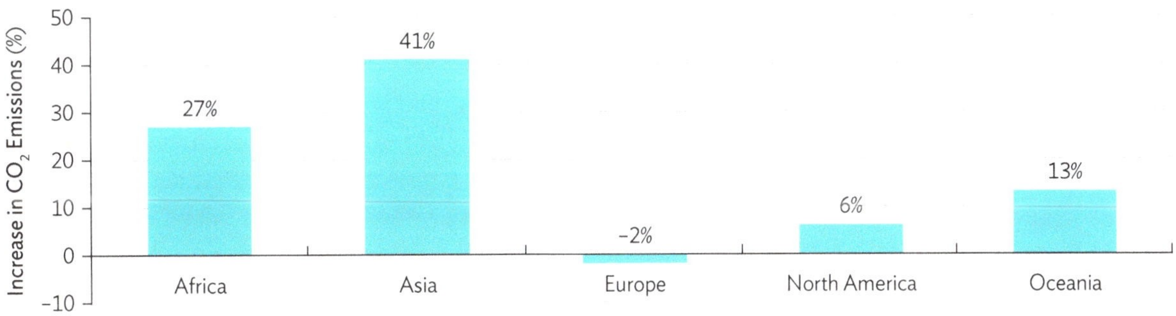

CO_2 = carbon dioxide.
Source: Crippa et al. (2020).

With the bulk of transport sector emissions coming from the road sector, urgent and impactful action is needed to decarbonize road traffic if countries are to deliver on the low-carbon future envisioned by the goals of the Paris Agreement on climate change and the 2030 Agenda for Sustainable Development. Within the road sector, most greenhouse gas (GHG) emissions come from the use of transport infrastructure (tailpipe emissions), rather than the provision of the infrastructure itself. However, as tailpipe emissions are impacted by the quality, design, and management of the road infrastructure provided, a holistic examination of options is required to yield the optimal solution that is consistent with low-carbon pathways for decarbonizing transport.

Appropriate tools are therefore needed to inform evidence-based transport investment decisions that will meaningfully support scaling up of low-carbon and sustainable transport investments.

Purpose

This publication presents a comprehensive framework and a tool for estimating GHG emissions associated with road sector investments. The tool is intended for use by road sector practitioners to assess the carbon emission impact of investment alternatives and inform decisions on appropriate avoid–shift–improve GHG reduction measures that align with the goals of the Paris Agreement on climate change and with sustainable transport development pathways.

"Avoid" measures refer to the need to avoid travel including through improved demand management. "Shift" refers to modal shifts toward low-emission transport and toward active mobility. "Improve" policies and measures focus on vehicle and fuel efficiency such as promoting uptake and mainstreaming of electric vehicles (EVs) and alternative low-carbon fuels.

The most significant opportunity to reduce whole life carbon emissions occurs during the planning stage, when project or program outcomes are being formulated and evaluated. The tool kit therefore should be utilized as early as possible in the project life cycle. This early use helps inform decisions on prioritizing decarbonization measures or policies that align with and support the transition to net-zero pathways as set out at the network, system, or national level.

The Approach

As preparation of ADB and other multilateral development banks' road investment projects typically include the use of the Highway Development and Management system (HDM-4) to inform economic appraisal of the investments, and as HDM-4 already contains vehicle tailpipe emissions model, it was determined that building a broad carbon emission framework around HDM-4 was appropriate. When the HDM-4 vehicle emission models were being developed in the early 1990s, the focus was on purely tailpipe (or exhaust) emissions—often referred to as "tank-to-wheel (TTW)" emissions. There was no consideration to properly including the implication of transitioning to low or zero emitting vehicles, embedded emissions in vehicles and road construction, or the emissions associated with the production and distribution of the fuel ("wheel-to-tank").

To this end, the approach discussed in this publication has been developed for the life cycle estimation of carbon emissions associated with road investment projects. The analysis framework is based around the post-analysis of HDM-4 outputs, to provide a robust estimate of the carbon emissions associated with both the road works and vehicle emissions. The methodology adopts a whole-of-life approach to GHG estimation by considering all key GHG sources including embedded emissions in vehicles, carbon footprint of road construction, emissions associated with the production of alternative fuels, and well-to-wheel (WTW) emissions for internal combustion engine (ICE) vehicles, hybrid vehicles, plug-in hybrid electric vehicles (PHEV), and battery electric vehicles (BEV).

The methodology has been incorporated into an Excel tool kit, enabling efficient and repeatable application of the approach. The tool can be applied together with HDM-4 to (i) investigate system-level GHG reduction for several avoid–shift–improve scenarios relative to a predefined base case or business-as-usual scenario, (ii) inform the screening of projects that aligns with low-carbon pathways at the planning stage, (iii) inform options that will reduce GHG emissions during project preparation, and (iv) include the social costs of carbon in the economic appraisal of road investment options.

Users should be familiar with the basic analysis framework and methodologies within the HDM-4 setup before using the tool kit, as it is essential that the HDM-4 setup is correctly configured and calibrated to gain reasonable results from the tool kit.

Findings and Recommendations

Several case studies on "typical" road investment projects and programs (refer to Section 3.2) have been run through the tool confirming that, for most road networks, the primary source of CO_2 emissions is from tailpipe emissions of ICE vehicles (more than 80%). As the transition to low- or zero-emission vehicles (LZEVs) occurs, the significance of the embedded emissions in the road will become more dominant.

The above findings naturally imply that the greatest opportunity to decarbonize road transport in the short term exists in tackling ICE vehicle emissions, such as scaling up electric mobility, eliminating congestion points, increasing active mobility and modal shift to public transport, traffic demand management, or some combination of these.

Finally, a range of road decarbonization policies and measures have been documented, and commentary on their potential usage has been provided. The viability and impact of the policies will naturally vary significantly from one geographic region to another—a policy that works well in one of Asia's megacities is unlikely to be ideal for a small Pacific island village (and vice versa). The intent of the tool is not to select the policy for the analyst, but rather to provide a robust and consistent methodology to analyze policies that the analyst comes up with.

METHODOLOGY

2

Overview

The analysis methodology takes the primary outputs from HDM-4 in terms of the annual work quantities (by work type), vehicle emissions (for internal combustion engine vehicles), vehicle-kilometers-traveled, energy required to run vehicles, and basic vehicle parameters—and combines these with user-provided constants such as emissions per vehicle manufactured, emissions from different types of physical works, and electricity grid factors to establish the overall carbon dioxide (CO_2) emissions (see Annex: The Tool Kit).

Inherent within the approach is the ability to competently conduct an HDM-4 analysis. The accuracy of HDM-4 modeling has a direct impact on the accuracy of CO_2 projections. This is particularly relevant when considering scenarios where traffic may divert onto the road under consideration, or where induced travel may occur. The requirement to accurately model such impacts within HDM-4 is a prerequisite for getting good results from the tool kit.

The methodology establishes an estimate of the total CO_2 for a given budget scenario in the case of a network-level analysis, or an investment alternative in the case of a project analysis within HDM-4.[1] The tool kit provides the ability to compare budget scenarios—for instance, with and without a project, or the impact of underfunding road maintenance and renewals on GHG emissions.

The methodology also incorporates the ability to analyze GHG trends associated with a transition away from internal combustion engines (ICEs) to low- or zero-emission vehicles (LZEVs) including electric vehicles (EVs). For EVs, it differentiates GHGs based on the source of power generation.

The approach breaks down the total CO_2 emissions into the following components:

- Traffic-based emissions, comprising
  - tailpipe emissions for ICE vehicles, including upstream well-to-tank (WTT) emissions in the production and transportation of fuels;
  - emissions from the generation of power for EVs; and
  - emissions associated with the manufacturing of vehicles.
- Road-based emissions, comprising
  - emissions from maintenance, renewal, and rehabilitation works and
  - emissions associated with specified capital improvement works.

HDM-4 includes models that predict tailpipe emissions from ICE vehicles; other sources of emissions illustrated in Figure 2 are covered in the tool kit.

1 A network analysis within HDM-4 typically examines how the overall condition of the network will change as a result of changes in the funding for routine maintenance, periodic resurfacing, and rehabilitation of the network. A project analysis within HDM-4 typically examines the impact of constructing a new route, adding capacity (more lanes) to an existing route or a similar type of scenarios.

Figure 2: Sources of Road-Related Emissions

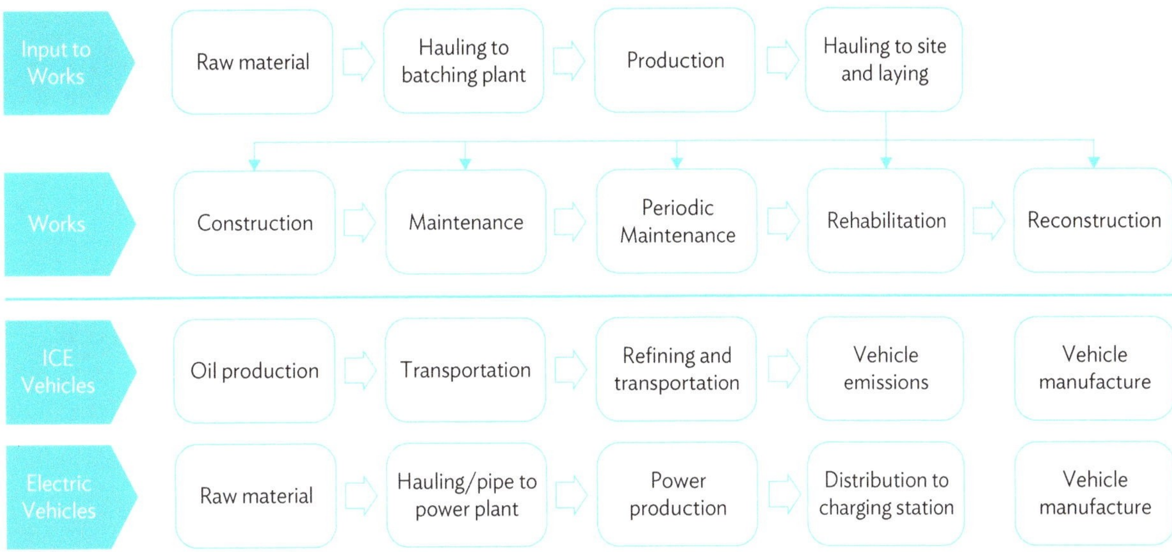

ICE = internal combustion engines.

Source: Authors.

Carbon Dioxide Equivalent

Emissions of non-CO_2 gases should be reflected in the inputs to the analysis of carbon dioxide equivalents (CO_2e), using the most recent Global Warming Potentials (GWPs) published by the Intergovernmental Panel on Climate Change (IPCC) and adopted by the United Nations Framework Convention on Climate Change (UNFCCC), based on 100-year time horizon.

Table 1 provides the 100-year period GWP values from the IPCC Sixth Assessment Report (AR6) for the three main GHGs alongside the values reported in previous assessment reports for comparison. Additional GWP for methane has been included to differentiate between methane that originates from fossil fuel source, and methane from nonfossil fuel sources, like agriculture.

Table 1: IPCC Sixth Assessment Report on Global Warming Potentials

Greenhouse Gas	AR4 2007	AR5 2014	AR6 2021
CO_2	1	1	1
$CH_{4\ \text{fossil origin}}$	N/A	30	29.8
$CH_{4\ \text{nonfossil origin}}$	25	28	27
N_2O	298	265	273

AR = assessment report, CH_4 = methane, CO_2 = carbon dioxide, IPCC = Intergovernmental Panel on Climate Change, N_2O = nitrous oxide.

Source: Intergovernmental Panel on Climate Change (2021).

The following relationship is used to derive CO_2e:

$$CO_2e = CO_2 + 29.8 \times CH_4 + 273 \times N_2O$$

Within HDM-4, it is only CO_2 that is predicted of these three emissions. Including methane and nitrous oxide into the calculation typically increases the overall value of CO_2e by around 1% to 3% above that of CO_2 alone.

Calculations

Each calculation is done on an annual basis (and sometimes on a vehicle basis) and then added up to yield the total value (however, the notation below has been kept simpler without showing these additional summations):

$$Total\_CO_2e = Veh\_CO_2e + Works\_CO_2e$$

Where:

$Total\_CO_2e$	is the total CO_2e in tons
Veh_CO_2e	is the CO_2e from vehicles in tons
$Works\_CO_2e$	is the CO_2e from road and other works in tons

1. Vehicle Emissions

Vehicle emissions comprise tailpipe emissions from ICE vehicles, emissions associated with usage of EVs, and vehicle manufacture emissions.[2] The calculation is as follows:

$$Veh\_CO_2e_{vy} = ICE\_CO_2e_{vy} + EV\_CO_2e_{vy} + Manu\_CO_2e_{vy}$$

Where:

$Veh\_CO_2e_{vy}$	is the total vehicle emission in tons
$ICE\_CO_2e_{vy}$	is the well-to-wheel emissions from ICE vehicles in tons
$EV\_CO_2e_{vy}$	is the emissions associated with EVs in tons
$Manu\_CO_2e_{vy}$	is the emissions associated with vehicle manufacture in tons
v	vehicle type
y	year

[2] For hybrid vehicles (including PHEVs), these should be modeled as ICEs but with revised fuel efficiency factors and increased manufacturing emissions to reflect the battery. HDM-5 is expected to model hybrid and EVs better than HDM-4, with improved ease and accuracy.

ICE_CO_2 is made up of two components:

(i) Tank-to-Wheel (TTW) emissions: These are directly estimated by HDM-4.

(ii) Well-to-Tank (WTT) emissions: This factor adjusts for emissions associated with the refining and distribution of fossil fuels.

$$ICE\_CO_2e_{vy} = WT \times TP\_CO_2e_{vy} \times pCO_2e_v/1000$$

Where:

$ICE\_CO_2e_{vys}$	are fuel emissions expressed in tons
WT	is the adjustment factor to account for the well-to-tank (≥ 1.0)
$TP\_CO_2e_{vy}$	is the predicted HDM-4 tailpipe emissions for ICE vehicles
pCO_2e_v	is the percentage of HDM-4 emissions to include for a vehicle type

$TP\_CO_2e_{vy}$ can be found by querying HDM-4 output database for the parameter EMISSQTY_4 in the Annual Data Table.[3]

For EVs, the emission quantities are calculated from the vehicle's energy usage data predicted by HDM-4. The CO_2e_{EV} is calculated by converting the energy usage from HDM-4 (in joules) to kilowatt-hours (kWh) and multiplying by the grid factor.

$$EV\_CO_2e_{vy} = MJ_{Vy} \times ECF \times Grid\_Factor$$

Where MJ_{Vy} is the total number of megajoule of energy usage for EVs from HDM-4.

MJ_{Vy} is available by querying the output HDM-4 database and by looking for parameter ENGFACT_12 within the Annual Vehicle Table.[4]

ECF is the factor for converting energy usage from HDM-4 (in joules) to kWh

$Grid\_Factor$ is the grid emission factor. It defines the CO_2e emissions produced by generating one kWh of electricity. Refer to Section 2.4.1.

The final component is the CO_2e associated with the production of the vehicles themselves, which is calculated on a vehicle type basis.[5] This is calculated by combining the CO_2 per vehicle, against the equivalent number of vehicles driven within the analysis.

$$Manu\_CO_2e_{vy} = \sum_{v=1 \text{ to } V}^{\text{all vehicles}} perVeh\_CO_2e_v \sum_{y=1 \text{ to } Y}^{\text{all years, by vehicle}} \frac{VKT_{vy}}{LIFEKM_{vy}}$$

3 The Annual Data Table is part of the HDM-4 output database. To access it, run an HDM-4 analysis, then open the output database in Microsoft Access.

4 The Annual Vehicle Table is part of the HDM-4 output database. To access it, run an HDM-4 analysis, then open the output database in Microsoft Access.

5 Within HDM-4, the user defines which vehicle types are to be included in the analysis. For example, a small passenger car is a vehicle type for which traffic projections are undertaken and all ensuing calculations (including in the tool kit) are based. HDM-4 does not constrain how many vehicle types are created, and the tool kit permits the transition of the vehicle fleet from ICE to LZEV over time.

Where $perVeh\_CO_2e_v$ is the estimated CO_2e for the production of a single vehicle in tons

VKT_{vy} is the HDM-4 calculated vehicle-km-traveled for a given vehicle in a given year

$LIFEKM_{vy}$ is the HDM-4 calculated expected life (in kilometer) for a given vehicle in a given year (the expected life within HDM-4 changes based on road condition)

V is the total number of vehicle types

Y is the total number of years in the analysis

VKT_{vy} is available by querying the output HDM-4 database and by looking for parameters AADTKC_0 from the Annual Vehicle Table and LENGTH from the Sections table (footnote 4). $LIFEKM_{vy}$ is available as parameter LIFEKM_K in the Annual Vehicle Table.

2. Road Works Emissions

Road works emissions CO_2e_{Works} is simply the multiplication of the unit rate of emissions (CO_2e per unit of treatment, for example, m[2]) against the HDM-4 provided annual quantities of work (by treatment type).

$$Works\_CO_2e_y = \sum_{\substack{y=1 \text{ to } Y \\ t=1 \text{ to } T}}^{\textit{All years, all Treatments}} Treat\_CO_2e_t \times Treat\_Quant_{ty}$$

Where $Treat\_CO_2e_t$ is the CO_2e per m^2 of treatment type t undertaken

$Treat\_Quant_{ty}$ is the quantity of treatment type t in year y from HDM-4

Estimating Non-HDM-4 Input Parameters

In addition to the need to calibrate HDM-4, there is also a need to estimate and provide inputs for the various parameters including grid factor, well-to-tank emission factor, typical emission associated with vehicle manufacture, road works emission factor, and the social costs of carbon that are fed into the post-HDM-4 analysis tool kit discussed in Part B. The actual value relevant to a particular analysis will require due consideration of the location and practices (including material sources) in use, among other considerations.

3. Grid Factor

The impact of EVs and their electricity consumption is related to the actual grid emissions, reflecting the emissions caused by EVs in the same manner as other electricity-consuming devices in the same country. The grid emission factor or grid factor can be derived from the total electricity produced minus losses and the total GHG emissions of the electricity production sector. Table 2 lists the grid factors for countries in Asia and the Pacific (the latest values should be used if available).

Table 2: Grid Factor by Country
($kgCO_2e/kWh$)

Country	Grid Factor	Year	Country	Grid Factor	Year
Armenia	0.16	2016	Nauru	0.92	2010
Azerbaijan	0.57	2015	Nepal	0	2015
Bhutan	0	2017	Pakistan	0.5	2015
Cambodia	0.7	2015	Palau	0.71	2011
PRC	0.79	2015	PNG	0.51	2007
Cook Islands	0.17	2017	Philippines	0.68	2015
Fiji	0.37	2015	Samoa	0.4	2007
Georgia	0.13	2015	Solomon Islands	0.94	2010
Indonesia	0.81	2015	Sri Lanka	0.55	2015
Kazakhstan	0.82	2015	Tajikistan	0.02	2015
Kiribati	0.71	2014	Thailand	0.55	2015
Kyrgyz Republic	0.28	2015	Timor-Leste	0.83	2014
Lao PDR	0	2017	Tonga	0.67	2008
Malaysia	0.73	2015	Turkmenistan	1.05	2015
Maldives	0.46	2012	Tuvalu	0.69	2015
Marshall Islands	0.68	2012	Uzbekistan	0.76	2015
FSM	0.69	2010	Vanuatu	0.55	2012
Mongolia	0.87	2015	Viet Nam	0.53	2015

PRC = People's Republic of China, Lao PDR = Lao People's Democratic Republic, FSM = Federated States of Micronesia, PNG = Papua New Guinea.

Source: Grutter and Kim (2019).

In countries with a high grid factor (typically 0.8 $kgCO_2/kWh$ or more), greening the grid should be a priority. The impact of EVs on GHG emission reduction will be lesser for these countries with a nongreen grid. Promoting transition to EVs or greening the grid in parallel to promotion of EVs is not considered an effective strategy since grid greening usually takes a long time because of long lifespan of energy production units (Grütter and Kim 2019).

4. Social Costs of Carbon

The calculation methodology embedded in the tool kit allows for unit social cost of carbon to be applied to estimated CO_2e to estimate the cost of benefit from the damage avoided with projects that reduce emissions and the cost of damage created with projects that increase emissions. For ADB projects, the social costs of carbon advised in the guidelines for economic analysis should be used (ADB 2017).[6] Table 3 shows that the current ADB social cost of carbon is at $57/ton of CO_2.

Table 3: ADB Social Cost of Carbon

Base Year	World Inflation	Social Cost of Carbon ($/ton of CO_2)
2016	–	36.3
2017	3.1	38.2
2018	2.7	40.0
2019	2.3	41.7
2020	1.6	43.2
2021	4.7	46.1
2022	7.2	50.4
2023	5.0	54.0
2024	3.5	57.0

ADB = Asian Development Bank, CO_2 = carbon dioxide.

Notes: For 2023–2024, world inflation was calculated using GDP deflators from IMF's World Economic Outlook Database, April 2024 (accessed 19 April 2024). For 2017–2022, world inflation was calculated using GDP deflators from previous versions of IMF's World Economic Outlook Database.

Source: ADB (2017).

5. Vehicle Manufacture Emissions

Emissions associated with the manufacture of vehicles arise from the production of the raw materials (steel, aluminum, rubber, and plastic in particular—plus battery materials for EVs), along with the energy usage associated with the actual molding and assembling of the materials into a vehicle. It is therefore necessary to consider both the sources of energy used in the assembly plant, along with the energy used in producing those raw materials.

6 ADB's Guidelines for Economic Analysis of Projects recommends social cost of carbon of $36.30 per ton of CO_2 or its equivalent in 2016 prices for 2016 emissions, to be increased by 2% annually in real terms.

Table 4 shows the typical embedded emissions (*perVeh_CO$_2$e$_v$*) in ICE, hybrid vehicles, plug-in hybrid vehicles (PHEVs), and EVs. While the embedded emissions in the Main Car (excluding batteries) for EVs is lower than those in ICEs, the overall embedded emissions in EVs is much higher, owing to the substantial emissions in the production of batteries (more than 40% of embedded emissions) (Patterson et al. 2011). The tool kit user will need to determine what is an appropriate value to use in their analysis based on a review of the vehicle fleet in the country.

Table 4: Vehicle Manufacture Emissions

		Embedded tCO$_2$e per unit			
Vehicle Description	**Type**	**Battery**	**Main Car (excluding battery)**	**Main Car (plus battery)**	**Source**
Average EU car	ICE	–	6.9	6.9	Carbon Brief (2020)
Standard ICE vehicle	ICE	–	5.6	5.6	Low Carbon Vehicle Partnership (2018)
Compact ICE vehicle	ICE	–	6.4	6.4	Brennan and Barder (2016)
Citroen C1	ICE	–	6	6	Berners-Lee (2010)
Renault Clio	ICE	–	6.8	6.8	Transport and Environment (2022)
Mid-sized ICE	ICE	–	7.3	7.3	Nealer (2015)
	ICE	–	6.4	6.4	Brennan and Barder (2016)
	ICE	–	6.2	6.2	PE International (2013)
Ford Mondeo	ICE	–	7.5	7.5	PE International (2013)
VW Golf	ICE	–	8.4	8.4	PE International (2013)
Full ICE	ICE	–	9.5	9.5	Nealer (2015)
VW Passat	ICE	–	9.8	9.8	Transport and Environment (2022)
Audi A7	ICE	–	11.8	11.8	Transport and Environment (2022)
SUVs	ICE	–	10–20	10–20	Berners-Lee and Clark (2010)
Trucks	ICE	–	30–100	30–100	Authors
Hybrid vehicle	HB	–	–	6.5	Low Carbon Vehicle Partnership (2018)
Prius Eco	HB	0.5	5.7	6.2	Carbon Brief (2020)
Plug-in hybrid vehicle	PHEV	–	–	6.7	Low Carbon Vehicle Partnership (2018)
Nissan Leaf 40 kWh (UK)	BEV	4.1	5.7	9.8	Carbon Brief (2020)
Nissan Leaf 40 kWh	BEV	–	–	9.2	Nealer (2015)
Nissan Leaf	BEV	–	–	10.5	Transport and Environment (2022)

continued on next page

Table 4 ***continued***

Vehicle Description	Type	Embedded tCO_2e per unit: Battery	Main Car (excluding battery)	Main Car (plus battery)	Source
Tesla model 3 (US batteries)	BEV	4.5	5.7	10.2	Carbon Brief (2020)
Tesla model 3 (Asian batteries)	BEV	7.5	5.7	13.2	Carbon Brief (2020)
Tesla model 3	BEV	–	–	12.6	Transport and Environment (2022)
Tesla model S 75 kWh	BEV	–	–	14.9	Nealer (2015)
Tesla model S	BEV	–	–	15.2	Transport and Environment (2022)
Battery Electric Vehicle	BEV	–	–	8.8	Low Carbon Vehicle Partnership (2018)
Compact BEV	BEV	–	–	15.9	Brennan and Barder (2016)
Mid-sized BEV	BEV	–	–	20.4	Brennan and Barder (2016)

BEV = battery electric vehicle, EU = European Union, HB = hybrid, ICE = internal combustion engine, PHEV = plug-in hybrid electric vehicle, SUV = sports utility vehicle, UK = United Kingdom, US = United States.

Source: Authors.

A report examining vehicle manufacturing in Europe by the Low Carbon Vehicle Partnership (2018) estimated the following factors:

- ICE – 5.6 tons/vehicle
- Hybrid – 6.5 tons/vehicle
- Plug-in hybrid – 6.7 tons/vehicle
- EV – 8.3 tons/vehicle

Meanwhile, Qiao et al. (2017) found that, owing to the much less efficient manufacturing techniques and limited recycling in the People's Republic of China at that time, emission rates for vehicle manufacture were much higher at:

- ICE – 9.2 tons/vehicle
- EV – 14.7 tons/vehicle

Following the analysis of some 790 vehicles, Buberger et al. (2022) estimated that the emissions from vehicle manufacture could be related to the weight of the vehicle body and battery size as follows:

$$\text{Manufacturing } CO_2e = 4.56 \times \text{Vehicle Weight} + 0.0835 \times \text{Battery Capacity}$$

where manufacturing CO_2e is in tons, vehicle weight is in tons and excludes the weight of batteries, and battery capacity is in kWh.

Table 5 presents a summary of CO_2e estimates from a range of sources for each of the HDM-4 vehicle classes, along with recommended default values for use in the tool kit using the work of Buberger et al. (2022). While values for EVs have only been provided for motorcycles and cars, the same approach can be applied to other vehicle classes as needed (including hybrid vehicles). More recent and accurate values of vehicle manufacture emissions should be used where available.

Table 5: Summary of CO_2e Estimates in HDM-4 Vehicle Classes

HDM-4 Vehicle Class	ICE/EV	Default Tare Weight (excluding batteries for EVs) (tons)	Battery Capacity for EVs (kWh)	Recommended Default (tons of CO_2e/vehicle)
Motorcycle	ICE	0.1		0.5
	EV	0.1	13	1.5
Small car	ICE	0.8		3.6
	EV	0.9	50	8.3
Medium car	ICE	1.0		4.6
	EV	1.1	50	9.2
Large car	ICE	1.2		5.5
	EV	1.3	75	12.2
Light delivery vehicle	ICE	1.3		5.9
Light goods vehicle	ICE	0.9		4.1
Four-wheel drive	ICE	1.5		6.8
Light truck	ICE	1.8		8.2
Medium truck	ICE	4.5		20.5
Heavy truck	ICE	9.0		41.0
Articulated truck	ICE	11.0		50.2
Minibus	ICE	1.1		5.0
Light bus	ICE	1.8		8.2
Medium bus	ICE	4.5		20.5
Heavy bus	ICE	8.0		36.5
Coach	ICE	10.0		45.6

CO_2e = carbon dioxide equivalent, EV = electric vehicle, HDM-4 = Highway Development and Management system, ICE = internal combustion engine, kWh = kilowatt-hour.

Source: Authors.

Within HDM-4, for the modeling of EVs, it is necessary to set the "Energy Used in Production" value at 0, otherwise the vehicle manufacturing energy will be double counted.

Figure 3: Energy Used in Production

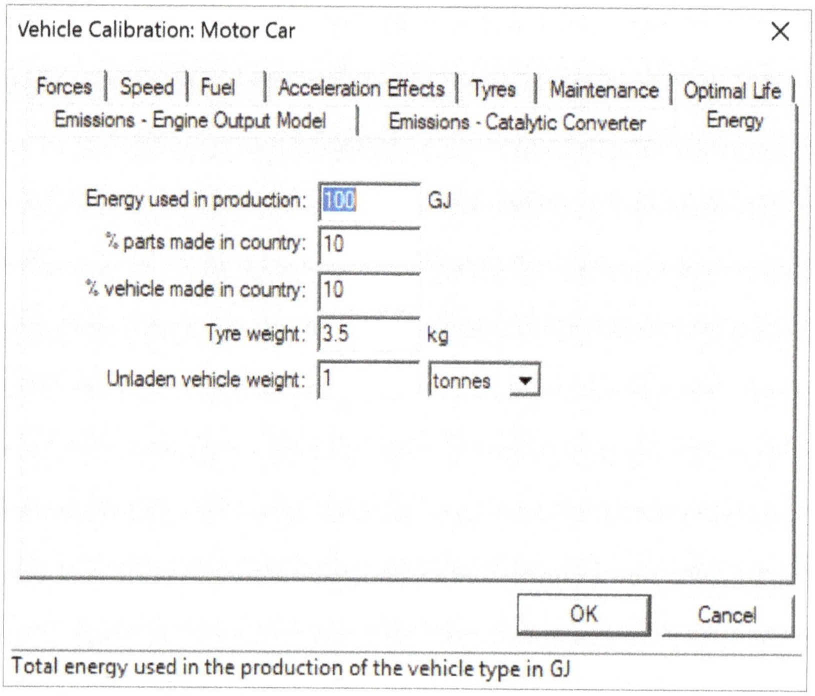

Source: HDM-4 software.

6. Road Works Emissions

Emissions during road works ($Treat\_CO_2e_y$) depend on several factors such as road design, types of construction materials, recycling of materials, the means used to transport the materials, distance traveled, and the type of vehicles used during construction. Suitable road works emission factors should be determined during project design. Table 6 shows the ranges for typical treatments.

Table 6: Road Works Emission Factors

Work Type	Unit Emission (CO_2e kg/m^2)
Edge repairs/potholes	20–40
Heavy patching	50–150
Thin AC overlay	20–40
Structural AC overlay	40–50
Full reconstruction	200–300

AC = Asphalt Concrete, CO_2e kg/m^2 = carbon dioxide equivalent kilogram per square meter.
Sources: Asian Development Bank (2010); World Bank and ASTAE (2010).

7. Well-to-Tank Emission Factor

HDM-4 calculates tank-to-wheel (TTW) emissions. To obtain well-to-wheel (WTW) emission estimates, Hoekstra (2020) recommends adjusting TTW emissions by a factor of 1.24 for diesel and 1.3 for petrol.

Offsetting Emissions

The analysis readily lends itself to an assessment of the number of trees needed to offset excess carbon emissions from road projects. Trees for Life (www.treesforlife.org.uk) estimates that six trees are needed to offset 1 ton of CO_2.

MetaMeta provides a much more detailed assessment of sequestration rates dependent upon the climatic conditions and tree planting arrangements.[7] The MetaMeta estimates are provided in Table 7.

Table 7: CO_2 Sequestration Rates from Planting Trees

Average CO_2 Sequestration for Different Planting Designs (Lifespan of 40 Years)				
		0–10 Years after Plantation (ton CO_2/km/year)	10–50 Years after Plantation (ton CO_2/km/year)	CO_2 Sequestration after 40 Years (ton CO_2/km)
Boreal climate	1 row of trees	2	4.2	145
	2 rows of trees	4.1	8.3	290
	4 rows of trees	8.2	16.6	580
	20 rows of trees	40.9	83.2	2,904
	Forest			100 ton CO_2/ha
Temperate climate	1 row of trees	3.1	6.3	220
	2 rows of trees	6.2	12.6	440
	4 rows of trees	12.4	25.2	880
	20 rows of trees	62	126	4,400
	Forest			320 ton CO_2/ha
Tropical climate	1 row of trees	3.9	7.9	275
	2 rows of trees	7.8	15.8	550
	4 rows of trees	15.5	31.5	1,100
	20 rows of trees	77.5	157	5,500
	Forest			400 ton CO_2/ha

CO_2 = carbon dioxide, ha = hectare, km = kilometer.
Source: MetaMeta.

7 ADB engaged MetaMeta of the Netherlands to produce the sequestration information for this study.

Typical Applications

The tool kit can be applied to integrate emission impacts of road investments and decarbonization policies during road investment analysis at project and network levels. It can be applied to inform road investment decisions and policies for countries striving for a low-carbon future (Figure 4 and Figure 5).

Figure 4: Options for Greener Road Infrastructure Development and Management

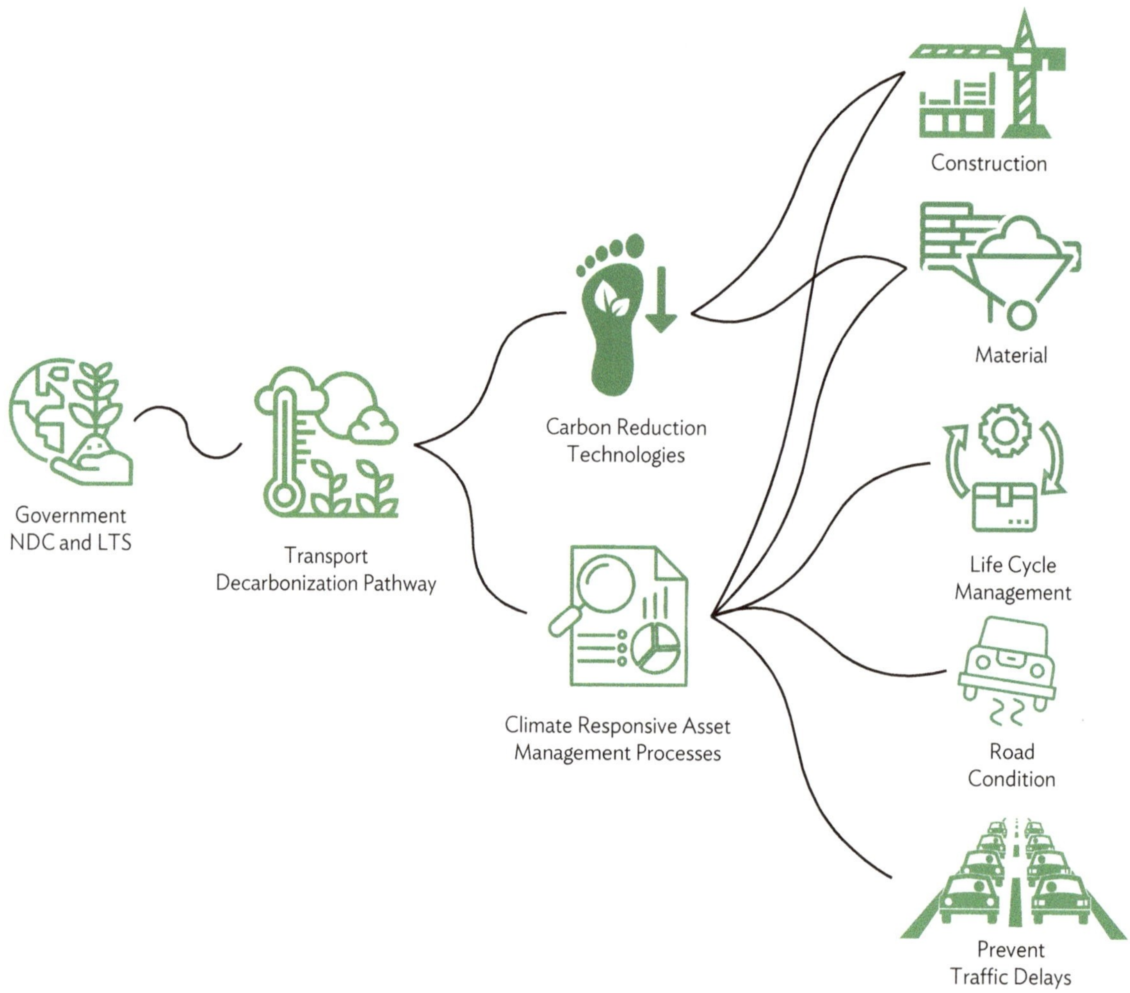

LTS = long-term strategy, NDC = Nationally Determined Contributions.

Source: Based upon National Treasury South Africa (2023).

Figure 5: Pathways to Low Vehicle Emissions

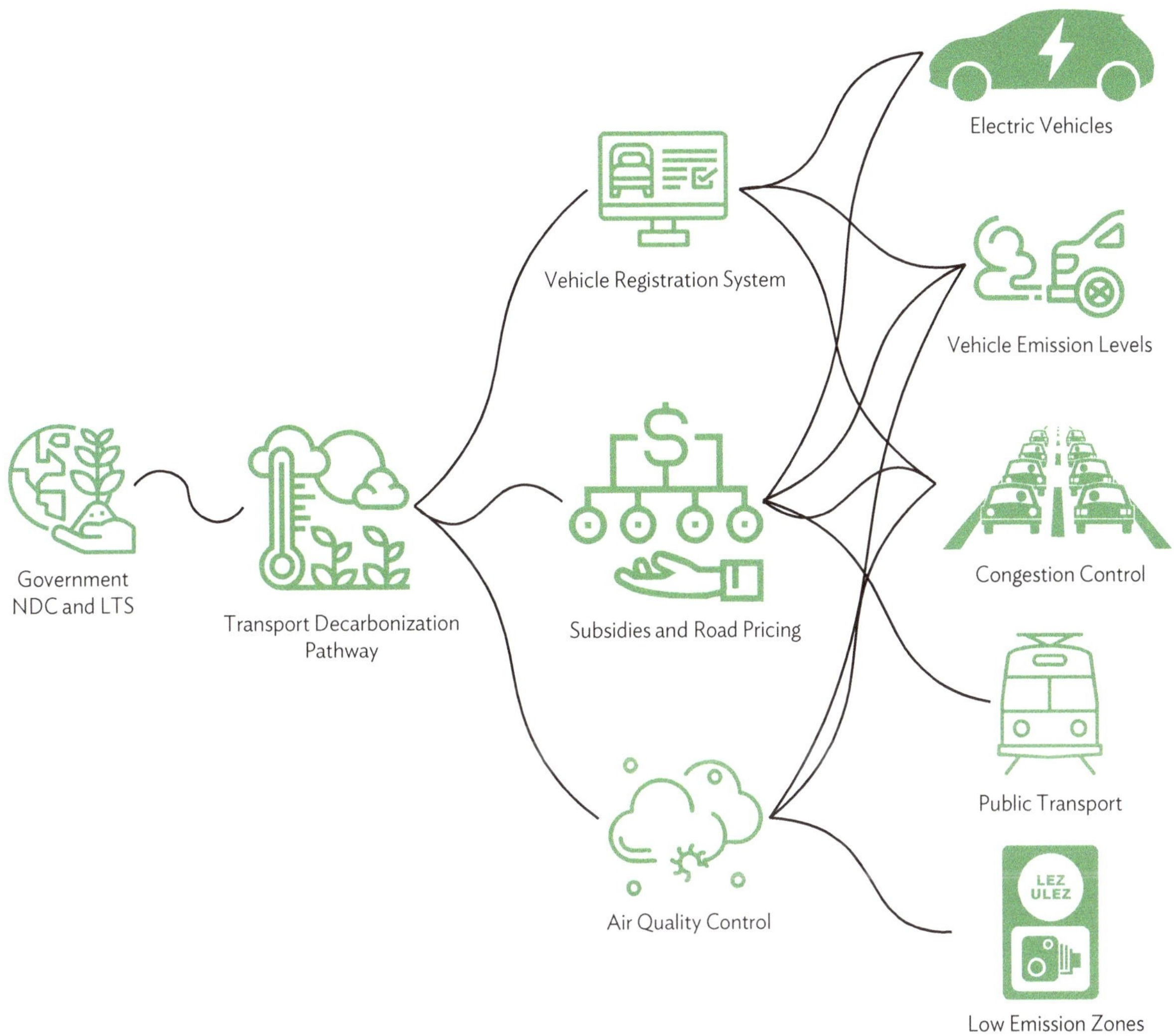

LTS = long-term strategy, NDC = Nationally Determined Contributions.

Source: Based upon Henning (2023).

The ADB Green Roads tool kit contains a range of measures to decarbonize transport systems, and will be updated regularly.[8]

Depending on the nature of the decarbonization analysis, it may require adjusting HDM-4, then re-running the analysis using the tool kit. Examples of different types of analysis, and the need to re-run HDM-4 to generate new outputs are given in Table 8.

[8] ADB Data Library. Green Roads Tool Kit.

Table 8: Example Types of Analysis and the Need to Re-Run HDM-4

Analysis Question	HDM-4 Impact
Impact of different budgets	☐ Can all be run at once, as HDM-4 will save all the different scenarios for subsequent carbon analysis. The tool kit will show the emission impacts of different budget levels.
Impact of changing intervention strategies	☐ Will need to run once, save the outputs (or complete the analysis), and then edit the intervention levels and re-run HDM-4 and the subsequent carbon analysis using the tool kit.
Impact of changing speed limits	☐ Will need to run once, save the outputs (or complete the analysis), then edit the road section data and re-run HDM-4 and the tool kit.
Impact of changing treatments to a lower carbon option	☐ If treatments have the same engineering and cost properties, then there is no need to re-run HDM-4. Otherwise, necessary edits need to be made and HDM-4 should be re-run.
Impact of policies to (i) move people to public transport, (ii) improve fuel and vehicle efficiencies, and (iii) reduce congestion	☐ Can change traffic data in HDM-4 and re-run the analysis. ☐ Can change percentage of HDM-4 emissions included in the tool kit from the input sheet of the tool kit without needing to rerun HDM-4.
Impact of policies to transition ICE vehicles to EVs and other forms of LZEVs	☐ Can change EV calibration factor in HDM-4 and rerun both HDM-4 and tool kit analysis. ☐ Can change grid factor and proportion of vehicles transitioning from ICE to EVs within the tool kit. This does not require re-running HDM-4.
Impact of additional investments and activities not included in HDM-4, such as charging infrastructure for e-mobility, ancillary road assets, and CO_2 sequestration (including deforestation).	☐ Various scenarios can be evaluated within the tool kit. It is not necessary to rerun HDM-4.

CO_2 = carbon dioxide, EV = electric vehicle, HDM-4 = Highway Development and Management system, ICE = internal combustion engine, LZEV = low- or zero-emission vehicle.

Source: Authors.

POLICIES AND CASE STUDIES 3

Road Decarbonization Policies

To effectively decarbonize road transport, a broad range of decarbonization policies must be implemented. The effectiveness of these policies will vary across different countries and regions, influenced by specific factors contributing to CO_2 emissions in the first place. Table 9 provides an overview of a range of potential road decarbonization policies and their applicability in reducing carbon emissions from road transport.

For most road networks, reducing emissions from vehicles using the roads can significantly impact overall CO_2 emission reduction, more so than addressing emissions from construction and maintenance activities. However, in rural areas with low traffic volumes, emissions from road construction and maintenance can constitute a substantial portion of the total emissions associated to the road network.

Table 9: Example Road Decarbonization Policies

Policy	Impact	Potential Usage
Policies applicable during construction of a new road or during rehabilitation activities		
Low-carbon road treatments	Reduce the CO_2/unit of physical works undertaken. Includes technologies such as recycling, incorporation of plastic wastes into asphalt concrete mix, and use of plastic in road surfacing.	All road networks (particularly low-traffic road networks where works-based CO_2 dominates total CO_2)
Constructing bypass of congested areas, including a grade separating major intersections	Enables non-congested travel to occur, with minimal speed change cycles. Will be an offset with higher works-based emissions from the extra road being built; but for congested areas, this would be more than an offset from the reduction in vehicle-based emissions. Options such as these have a high potential to induce traffic. Where possible, limiting use of additional capacity to LZEVs or high occupancy vehicles should be considered.	All congested areas
Policies applicable during operation and maintenance of roads		
Maintaining roads in a good to fair state of repair	Minimizes the quantity of reactive repair works (e.g., pothole filling) and reduces rolling resistance of vehicles (lowers fuel consumption).	All road networks

continued on next page

Table 9 ***continued***

Policy	Impact	Potential Usage
Optimizing work zones—duration and timing of closures	Minimizes traffic congestion impacts, including using lane rental schemes or similar.	Higher trafficked roads
Implementing overloading controls	Reduces fuel consumption for heavy vehicles and reduces excessive wear on road pavements (hence reducing frequency of works).	All road networks
Addressing transport logistics This includes approaches such as building logistics and distribution hubs at outer fringe of congested areas and use of EVs for short trips including last mile deliveries, improving border crossing logistics, and implanting free-flow tolling	Reduces emissions within the congested areas.	All road networks
Policies applicable to vehicles and traffic demand		
Managing vehicle speeds to minimize speed change cycles (traffic signals, intelligent transportation systems, congestion impacts, etc.) and to avoid excessively high average speeds	Speed changes result in additional fuel consumed in comparison to steady speed travel. Fuel consumption forms a u-shaped curve with speed, such that avoiding very high speeds or very low speeds is important for minimizing fuel consumption and vehicle emissions. Typically, speeds of about 70–80 kilometer/hour are optimal, but this varies significantly by vehicle type, terrain, and other factors.	All road networks
Encouraging transition to public transport or rail transport	Reduces private vehicles but increases buses on the road. However, if bus stops are not well-designed, congestion for private vehicles may worsen. If bus lanes are provided, then these would result in additional CO_2 emissions associated with their construction and maintenance.	Where demand on a route is high

continued on next page

Table 9 *continued*

Policy	Impact	Potential Usage
Encouraging transition to low- or zero-emission vehicles and fuels through a combination of subsidies, improved charging infrastructure, priority lanes, or similar Setting standards regarding vehicle emission for new and old vehicles (i.e., requirement for low emission vehicles and/or removal of older vehicles)	Eliminates the most polluting vehicles from the network and replaces them with more efficient vehicles—either low emission ICE or EV/hybrid vehicles. Eliminates tailpipe emissions and replaces them with power generation emissions. However, as EVs have significantly higher manufacturing CO_2 levels than ICEs, the benefit may not be realized in areas with very low vehicle usage. The life cycle emission of EVs is projected to significantly fall as renewable energy sources become more mainstream.	Where power generation is "cleaner" than equivalent ICE running, and where vehicle usage enables a reduction in life-cycle CO_2 emissions
Tolling, increased fuel taxes, road user charging, or similar	Decreases private vehicle usage but requires availability of alternative modes of travel.	Where alternative modes of travel exist
Improving emission standards on vehicles and power generation ("greening the grid")	Reduces CO_2 emissions from vehicles and from the charging of EVs.	All networks

CO_2 = carbon dioxide, EV = electric vehicle, ICE = internal combustion engine, LZEV = low- or zero-emission vehicle.
Source: Authors.

For any policy, there is a need to also consider whether there will be a transfer of vehicles onto different roads, or whether the policy would induce additional travel. All such considerations should be included in the HDM-4 setup to ensure that the correct traffic volumes are used in the analysis, and that the full impacts of any policy are captured. The policies have been broadly categorized into those applicable to the construction and rehabilitation of roads; those applicable during the operation and maintenance of roads; and those that are vehicle based—although there is cross-over in many cases.

Case Studies

The tool kit includes case studies ranging from high-volume congested road networks in Asia, to low-volume roads on a Pacific Island. The case studies illustrate the impact of different levels of investment for maintenance and renewals at a network level, project based four-laning of a congested road, and policies to transition from petrol to electric motorcycles.

Case Study 1: Upgrading Existing Two-Lane Paved Road to Four Lanes

Country Region	Central and West Asia				
Project Type	Highway improvement				
Objectives	Project analysis on the impact of expanding capacity from two lanes to four lanes. Note that this is from a project where a detailed study had been undertaken on the expected traffic flows (including induced traffic) once the project was completed. These flows were then included into the HDM-4 analysis, and thus incorporated into the tool kit.				
Total Road Length	560 km				
Vehicle Type	9				
Analysis Period	23 years				
Base Case	Retain road as two lanes				
Option(s)	Expand road to four lanes with divided median				
Traffic	Base case has a traffic growth at 2% per year Option has a traffic growth at 4% per year				
Alternatives	Base Case		Option		Net
Units	tCO_2e	%	tCO_2e	%	tCO_2e
Works	642,407	1.2	6,161,585	9.9	5,519,178
ICE Vehicles	47,814,917	86	49,134,547	79	1,319,630
EV Vehicles	0	0	0	0	0
Vehicle Manufacture	6,916,640	12	6,997,115	11	80,475
Total	**55,373,963**	**100**	**62,293,248**	**100**	**6,919,285**

continued on next page

Case Study 1 ***continued***

Base – Tons of CO_2e Emissions by Year

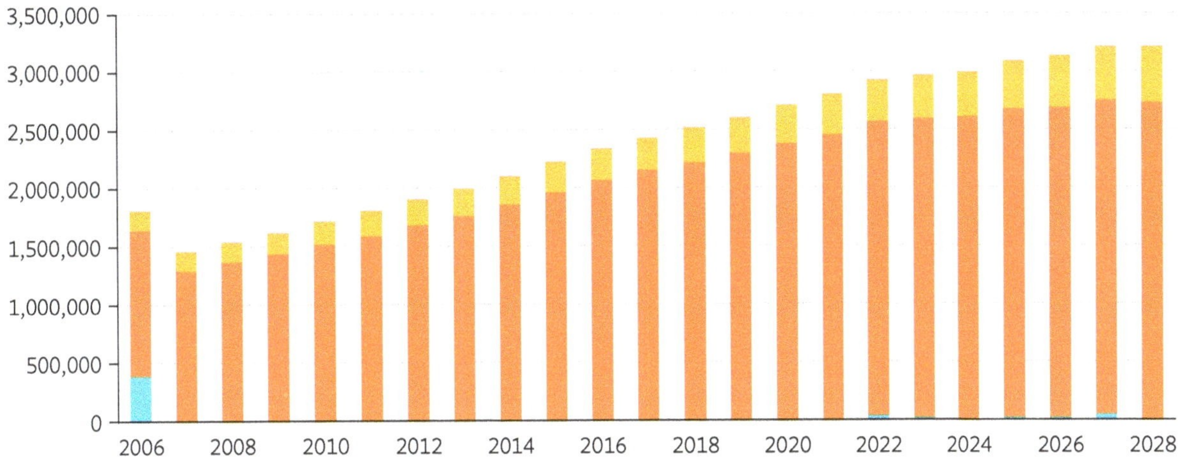

Option – Tons of CO_2e Emissions by Year

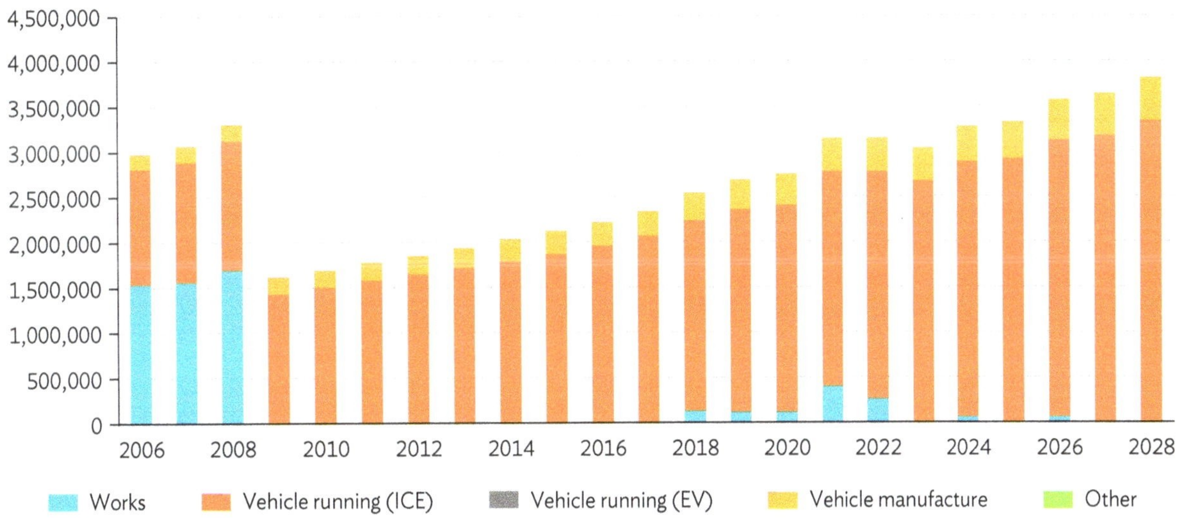

Mitigation Actions	The following mitigation actions were considered in the project design: (i) using recycled materials during construction, (ii) optimizing material sources to reduce GHG emissions, (iii) implementing axle load controls to minimize overloading, (iv) reducing speed limit from 100 km/h to 80 km/h, and (v) tree planting for CO_2 sequestration.

CO_2 = carbon dioxide, EV = electric vehicle, GHG = greenhouse gas, h = hour, HDM-4 = Highway Development and Management system, ICE = internal combustion engine, km = kilometer, tCO_2e = tons of carbon dioxide equivalent.

Source: Authors.

Case Study 2: Impact of Network Budget Allocations and EV Transition on Emissions and Economic Benefits (Net Present Value)

Country Region	Central and West Asia				
Objectives	Strategy analysis to demonstrate (i) the impact of different funding levels on emissions and net present value (NPV) and (ii) the impact of transition to EVs.				
Network Length	18,700 km				
Traffic Distribution	Distribution of AADT on Public Road Network (see chart below)				
Analysis Period	31 years				
Base Case	**Business as usual (BAU)** – based on pre-pandemic (2019) network expenditure of $25 million per year or $1,322 per km per year.				
Option(s)	100%, 80%, 60%, and 40% of unconstrained budget (UB)				
Budget Scenarios	BAU	40% UB[a]	60% UB	80% UB	100% UB
Units	Million Tons tCO_2e				
Works	2.69	6.54	9.40	12.26	17.89
ICE Vehicles	310.78	310.81	303.75	301.16	294.74
EV Vehicles	0	0	0	0	0
Vehicle Manufacture	57.41	51.92	48.50	46.84	44.23
Total	**370.88**	**369.27**	**361.64**	**360.26**	**356.86**

Distribution of AADT on Public Road Network

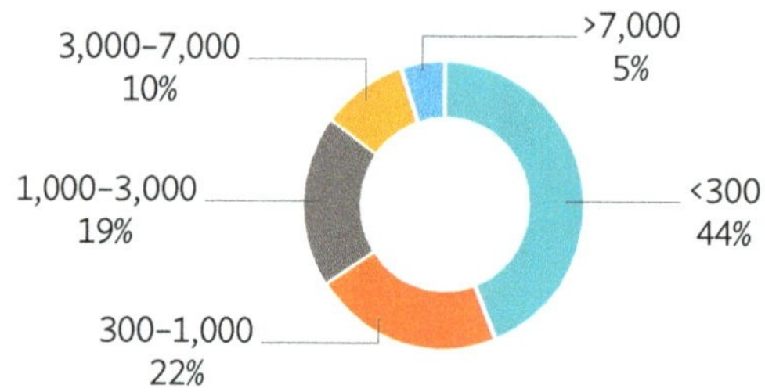

Investment Needs Without Consideration of Low-Carbon Transition

Budget Scenarios	Road Network Investment Needs[b] 31-Year Total (Mio $)	Annual (Mio $ per year)	%GDP	$/Year/km	Average Condition[c] (IRI m/km)	Emissions (Mio Ton. CO_2e)	Total Transport Costs[d] (Mio $)	NPV[e] (Mio $)	Incremental BCR
SC1: Business as Usual	765.60	24.7	0.33	1,321.60	8.8	370.9	55,694.30	–	1.0
SC2: Budget @ 60% less optimal	2,225.30	71.8	0.96	3,841.60	6.9	369.3	53,261.50	2,432.80	5.0
SC3: Budget @ 40% less optimal	3,338.00	107.7	1.44	5,762.50	5.9	361.6	51,239.60	4,454.80	5.1
SC4: Budget @ 20% less optimal	4,450.70	143.6	1.92	7,683.30	5.1	360.3	50,514.90	5,179.50	4.3
SC5: Unconstrained (optimal budget)	5,563.30	179.5	2.40	9,604.10	4.0	356.9	49,414.60	6,279.70	4.3

continued on next page

Case Study 2 *continued*

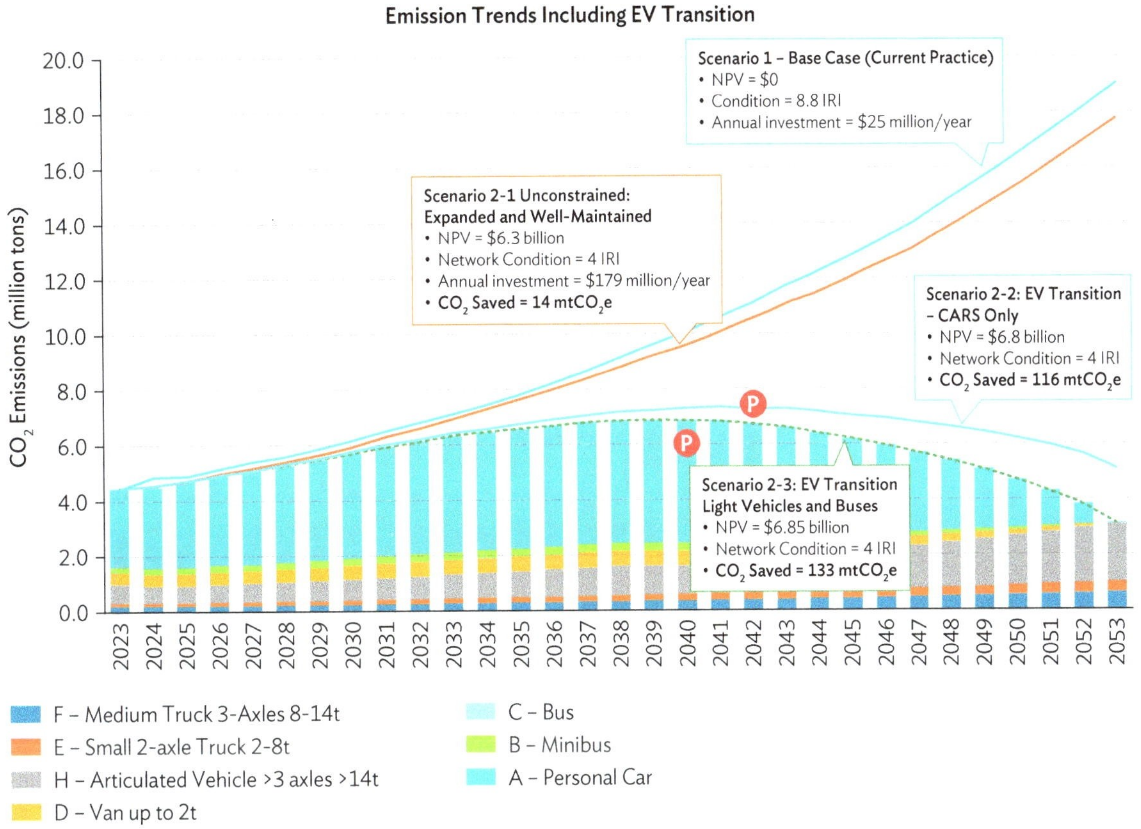

Mitigation Actions	The investment needs table and emission trend graph above demonstrate reduced emissions with improved road condition. However, much larger contribution to decarbonizing the road network would be achieved by introducing EVs. The results also demonstrate that, with the appropriate pathway to decarbonization, it is economically feasible for countries to continue investing in expanding and maintaining their road networks and concurrently decarbonize road transport. Alternative sources of funding for the road network (and in many cases, general revenue) will be required as countries move away from ICEs toward LZEVs and as fuel taxes taper off.

AADT = annual average daily traffic, EV = electric vehicle, ICE = internal combustion engine, km = kilometer, LZEV = low- or zero-emission vehicle.

[a] A 40% unconstrained budget denotes 40% of the unconstrained budget totaling $5.56 billion over 31 years. Unconstrained budget is the allocation needed to eliminate backlogs (roads in poor and very poor condition) and maintain the network in good condition over the analysis period.

[b] Road Network Investment Needs: Refers to the total expenditures required by road agencies for the rehabilitation and maintenance of the road network. These are commonly known as Road Agency Costs (RAC).

[c] Average Condition: Represents the mean value of the International Roughness Index (IRI) across the entire road network, serving as an indicator of overall pavement smoothness and ride quality.

[d] Total Transport Costs: The combined cost of Road Agency Costs (RAC) and Road User Costs (RUC), reflecting the full economic burden of road transport, including both infrastructure and user-related expenses.

[e] NPV (Net Present Value): The incremental Net Present Value calculated for each scenario, defined as the difference in total transport costs between the given scenario and Scenario 1. It indicates the relative economic efficiency of each investment strategy.

Source: Authors.

Case Study 3: Upgrading Existing Earth Road to Two-Lane Paved Road

Country Region	East Asia				
Type of Project	Road upgrading – earth road to paved road				
Objectives	Project analysis to examine the impact of upgrading a low-volume rural road from earth to paved				
Total Road Length	200 km				
Vehicle Types	7				
Analysis Period	22 years				
Base Case	Retain road as earth/gravel road, with grading twice a year and annual spot regravelling at selected locations.				
Option(s)	Upgrade to a paved road surface over a 3-year period, with resurfacing 14 years later.				
Traffic	Base annual average daily traffic of 75 vehicles per day, with a growth rate of 5%. No induced traffic considered because a remote location and a lack of alternative routes and modes of transport suggest that trips are limited and/or suppressed.				
	100%, 80%, 60%, and 40% of unconstrained budget (UB)				
Alternatives	Base Case		Option		Net
Units	tCO_2e	%	tCO_2e	%	tCO_2e
Works	33,106	56.3	133,944	84.2	100,838
ICE Vehicles	20,476	34.8	20,370	12.8	–106
EV Vehicles	0	0.0	0	0.0	0
Vehicle Manufacture	5,265	8.9	4,780	3.0	–485
Total	**58,847**	**100.0**	**159,094**	**100.0**	**100,247**

Base - Tons of CO_2e Emissions by Year

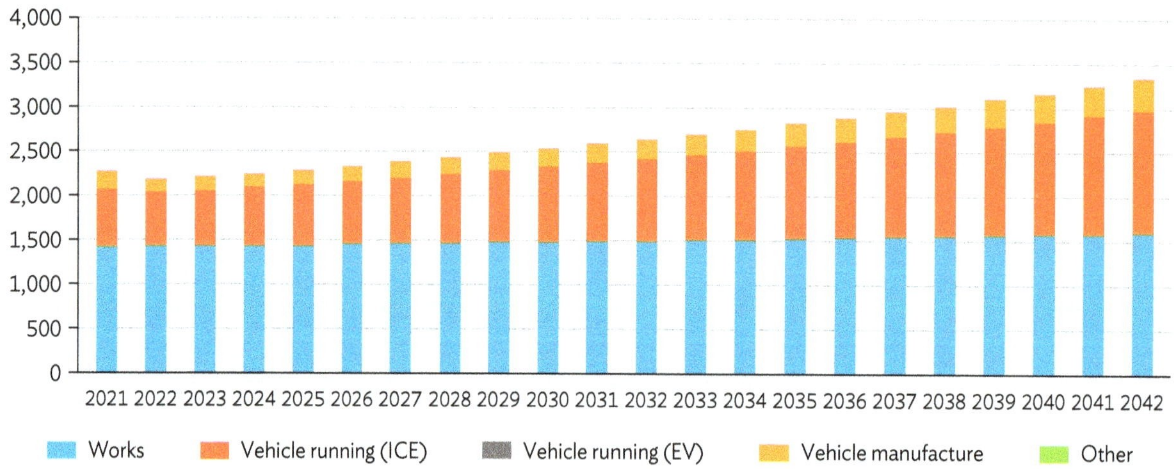

continued on next page

Case Study 3 *continued*

Option – Tons of CO_2e Emissions by Year

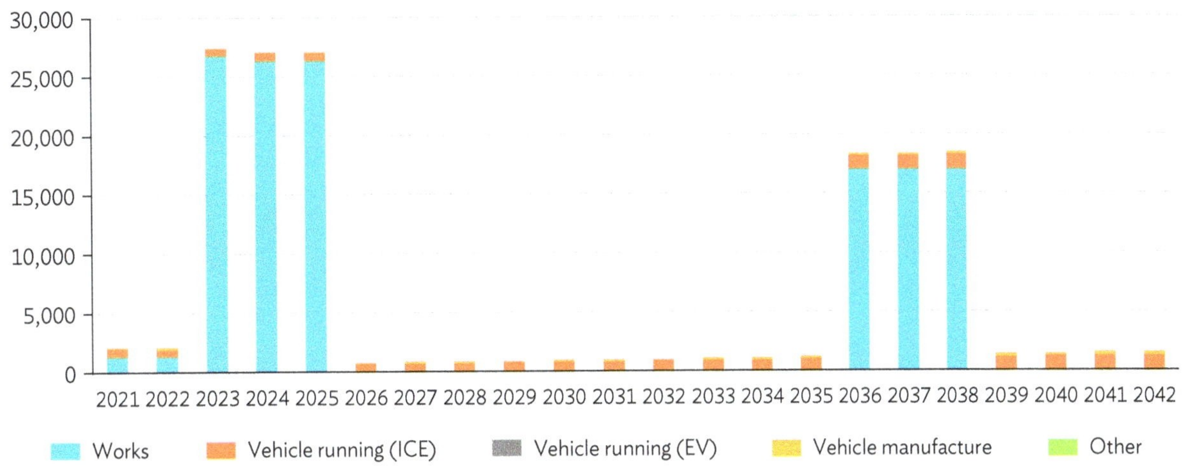

Remarks	The proposed mitigation actions considered were (i) tree planting for CO_2 sequestration in collaboration with the government's tree planting program and (ii) seeking the government's agreement on a policy measure to decarbonize transport during project preparation.

EV = electric vehicle, ICE = internal combustion engine, km = kilometer, tCO_2e = tons of carbon dioxide equivalent.

Source: Authors.

Case Study 4: Impact of Transition from Private Cars to Public Transport

Country Region	South Asia
Objective	Examine the impacts of policy for users transitioning from cars to buses or trains.
Type of Analysis	Strategy (network level)
Total Road Length	17,906 km

Traffic Distribution and Vehicle Composition

Distribution of AADT on the Road Network

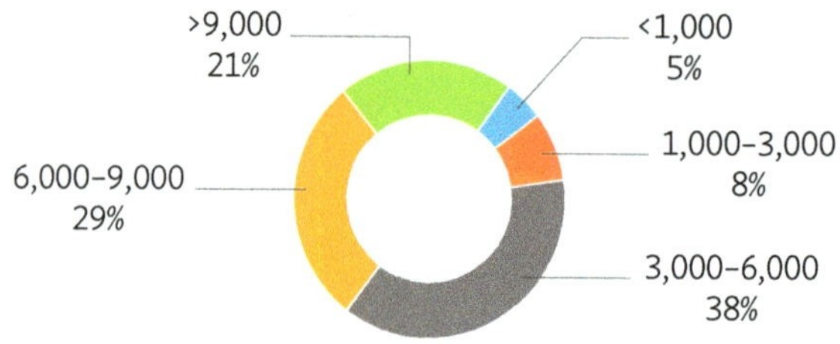

Road Network Traffic Composition

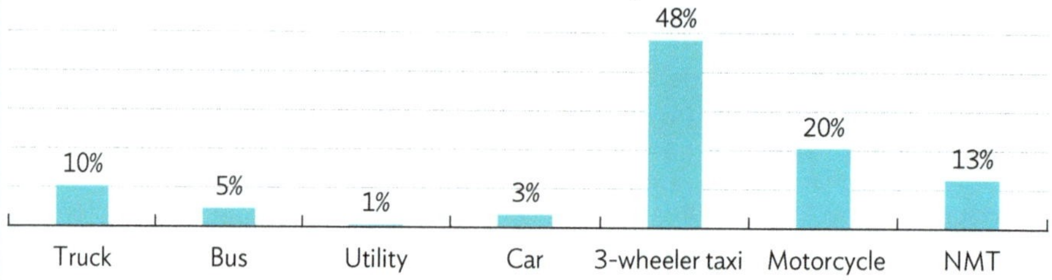

Analysis Period	25 years
Base Case	**Option 0:** No transition from private cars to public transport
Option(s)	**Option 1:** To Bus = 20% transition from private cars to buses over a period of 10 years. Results in additional buses on the network. **Option 2:** To Train = 20% transition from private cars to trains over a period of 10 years.

$mtCO_2e$ by Option and Source	Base Case	Transition to Buses		Transition to Trains	
	$mtCO_2e$	$mtCO_2e$	Net $mtCO_2e$	Net $mtCO_2e$	Net $mtCO_2e$
Works	9.074	9.042	−0.032	9.042	−0.032
ICE Vehicles	766.828	760.979	−5.849	760.710	−6.119
EV Vehicles	0	0	0	0	0
Vehicle Manufacture	79.728	79.534	−0.194	79.525	−0.202
Total	**855.630**	**849.554**	**−6.076**	**849.277**	**−6.353**

Remarks	This case study assumed a bus would be added to the road network for every four cars that have been removed due to the modal shift. Up to 6 million tons of emissions would be eliminated because of transition to buses, with slightly larger reduction (6.3 million tons) with transition to trains.

AADT = annual average daily traffic, EV = electric vehicle, ICE = internal combustion engine, km = kilometer, NMT = nonmotorized transport, tCO_2e = tons of carbon dioxide equivalent.

Source: Authors.

Case Study 5: Impact of Policy for Transitioning from ICE Motorbikes to Electric Motorbikes

Country Region	South Asia
Objective	Examine the impacts of policy for transitioning from ICE motorbikes to e-motorbikes
Type of Analysis	Strategy
Network Length	17,906 km
Vehicle Composition	(see chart below)

Vehicle Composition

Road Network Traffic Composition

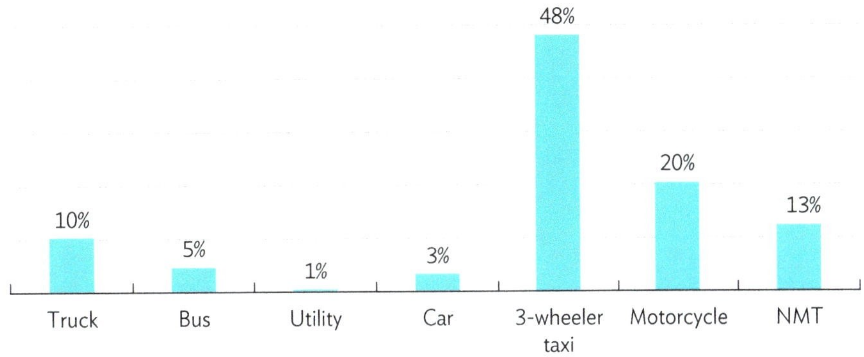

Analysis Period	25 years
Base Case	No transition to e-motorbikes
Option(s)	10% transition per year for 10 years (i.e., complete transition over a 10-year period), assuming emissions from e-motorbikes are 20% of the emissions from ICE motorbikes.

tCO_2e	Base Case	Option	Net
Works	9,074,207	9,074,207	0
ICE Vehicles	766,828,168	748,826,979	–18,001,189
EV Vehicles	0	3,600,238	3,600,238
Vehicle Manufacture	79,727,938	79,727,938	0
Total	**855,630,313**	**841,229,362**	**–14,400,952**

Remarks

An estimated reduction of 14 $mtCO_2e$ would be achieved with a 10% transition from ICE motorcycles to electric motorbikes. Considering 70% of the vehicle fleet comprises motorized two- or three-wheelers (see vehicle composition above), a much larger reduction in emission would be realized by replacing existing motorized two- or three-wheelers with electric versions.

In South Asia, motorized two- or three-wheeler road users are among the most vulnerable to serious injuries and death from road crashes. Significant co-benefits can be achieved by ensuring safer electric two- or three-wheelers, and by providing dedicated lanes in high risk areas in addition to other road safety interventions.

EV = electric vehicle, ICE = internal combustion engine, km = kilometer, NMT = nonmotorized transport, tCO_2e = tons of carbon dioxide equivalent.

Source: Authors.

Case Study 6: Impact of Budget Levels on GHG Emissions for a Relatively Low Traffic Volume Road Network

Country Region	Pacific (small island states)				
Objective	Examine the impacts of differing levels of maintenance on network CO_2 emissions				
Type of Analysis	Strategy analysis based on network level model for paved roads. Split by island (2), traffic bands (3), and initial road condition (3)				
Total Road Length	867 km				
Total Number of Vehicle Types	9				
Analysis Period	20 years				
Base Case	Equivalent of 40% of unconstrained budget				
Option(s)	Unconstrained budget and 10%, 20%, or 30% budget cuts				
$mtCO_2e$	Base Case	Unconstrained	10% Reduction	20% Reduction	30% Reduction
Works	0.446	0.757	0.732	0.701	0.571
ICE Vehicles	1.431	1.416	1.419	1.424	1.426
EV Vehicles	0	0	0	0	0
Vehicle Manufacture	0.306	0.289	0.292	0.297	0.302
Total	**2.182**	**2.462**	**2.442**	**2.422**	**2.299**

Road Network CO_2e Emission by Budget Levels

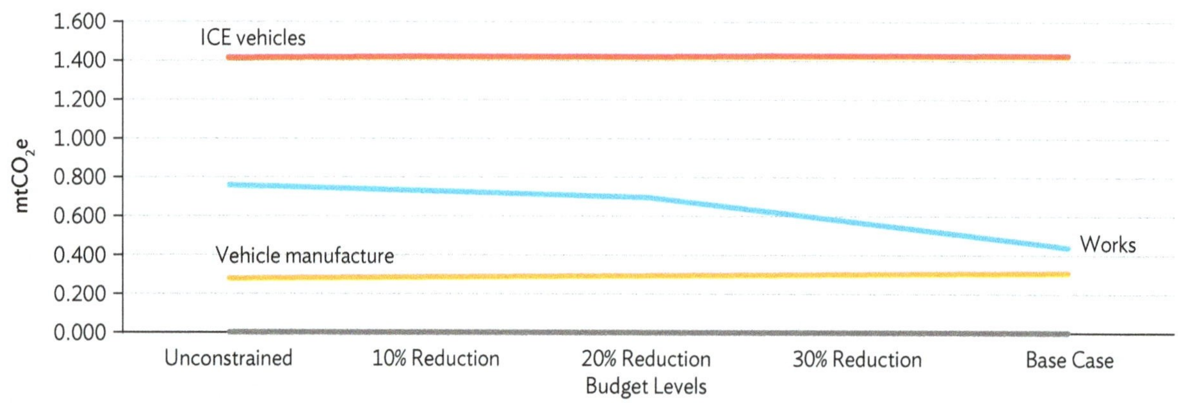

Remarks	Due to low traffic volume and low traffic growth, the increase in ICE vehicle and vehicle manufacture emissions as roads budget are reduced is outweighed by the estimated large decrease in works emission. Consequently, for this relatively "low traffic" road network, policies and measures that promote low-carbon construction and maintenance at scale, such as recycling of construction materials, will lead to significant GHG emission reduction relative to the base case scenario.

EV = electric vehicle, ICE = internal combustion engine, km = kilometer, $mtCO_2e$ = metric tons of carbon dioxide equivalent.

Source: Authors.

ANNEX: THE TOOL KIT

This annex has been prepared to align with version "v09_021" of the tool kit, which is the latest version as of 1 July 2024.

A. HOW TO USE THE TOOL KIT

A.1 About the Tool Kit

This section describes how to use the tool kit in estimating the greenhouse gas (GHG) emissions associated with road investment strategies and policies. The tool kit can be used alongside Highway Development and Management Model 4 (HDM-4) to

- assess the impact of different levels of road investment funding on GHG emissions;
- include social costs of carbon in the economic appraisal of road investment options; and
- demonstrate the economic feasibility of investing in roads to improve access (SDG 9.1.1), strengthen road infrastructure resilience and adaptive capacity to climate change and disasters triggered by natural hazards (SDG 13.1), and improve road infrastructure safety (SDG 3.6) while implementing policies and measures to decarbonize the road transport in a manner that aligns with transport emission reduction pathways and strategies of countries.

A.2 Tool Kit Structure

The tool kit is made up of several sheets which are all color coded:

Homesheet

The home sheet contains some basic instructions and the version history of the tool kit. Please ensure you are always using the latest version of the tool kit.

Input Tables

The input data and tables are entered into this sheet.

Scenarios

The scenario sheet enables you to select the options from HDM-4 to analyze and provide some additional inputs that are not by default computed in HDM-4.

Analysis-HDM4WorksEmissions

The analysis sheets contain the detailed calculations performed by the tool kit.

Output-TotalEmissionGraphs

The output sheets present the analysis results for up to five scenarios defined by the user.

The reference sheets contain some default data that may be useful for users setting up their analysis in the tool kit for the first time.

Cells in the tool kit that can be modified by a user are indicated by an orange cell:

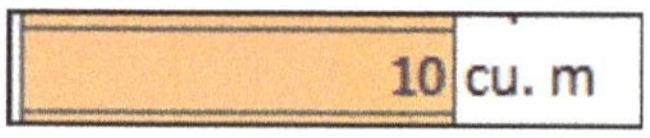

Some input cells would require you to select from a drop-down list. These are indicated by an arrow appearing when the user clicks within a cell:

Clicking on the down arrow will display the list:

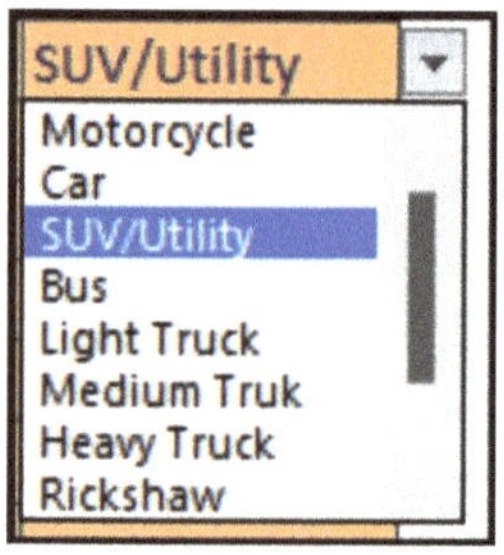

A.3 Navigating the Tool Kit

To quickly navigate to the sheets, a drop-down list of the sheets is displayed in the header of each sheet. Selecting the required sheet will automatically take you to that sheet.

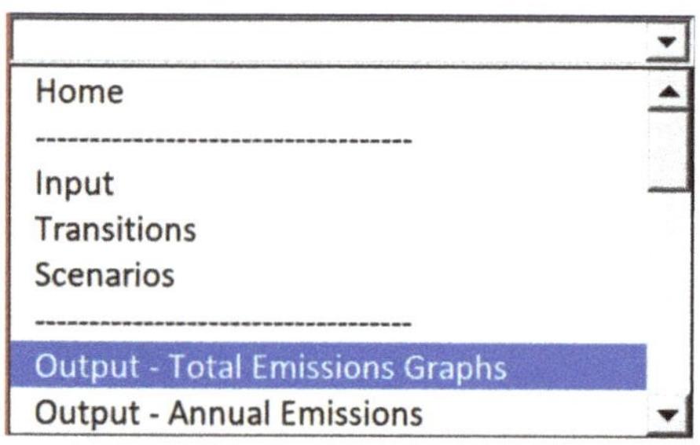

Tool kit navigation option 1

Alternatively, you can use the tabs displayed at the bottom of the worksheet, noting that you may have to use the scrollbar to locate some of the worksheets.

Tool kit navigation option 2

A.4 Considerations for Using the Tool Kit

The tool kit is designed to work alongside HDM-4 version 2. An analysis must first be performed in HDM-4 prior to using the tool kit. The tool kit supports the following analysis modes from HDM-4:

- Project analysis: analysis by project mode must be selected.
- Program and strategy analysis with budget scenario analysis completed.

Note that when performing an analysis with HDM-4, the following restrictions apply:

Analysis duration: 50 years maximum

Number of work items: limited to 100 individual work items and/or improvements

Vehicles: 50 individual vehicle types

When running an analysis within HDM-4, ensure that the following selections are made:

- If Fleet Transitions are to be used, then set the “Energy used in production” to 0 GJ for all vehicles within the vehicle fleet (this will avoid double counting within the tool kit).

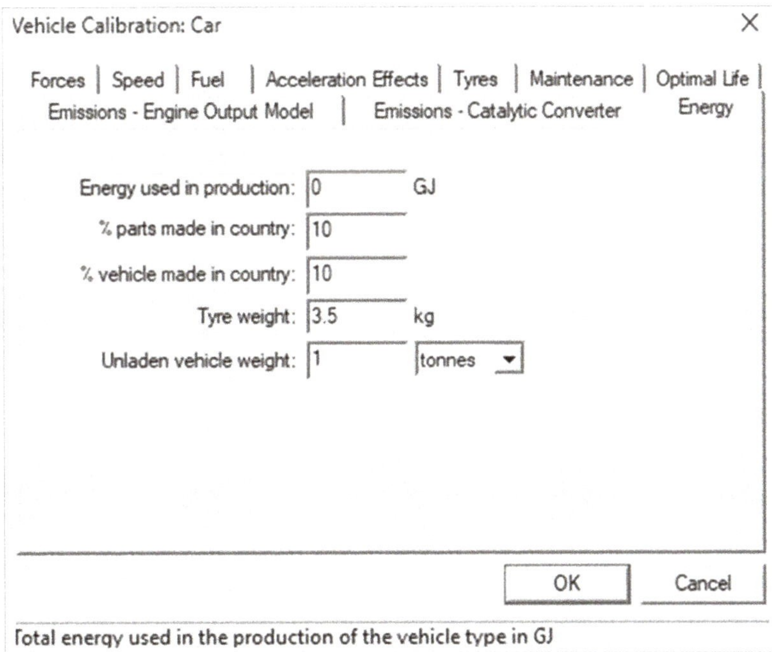

▲ **Energy used in production**

- Model Inclusion – Energy Balance
- Model Inclusion – Emissions
- Run Data Export Detail – untick “Exclude annual vehicle data”

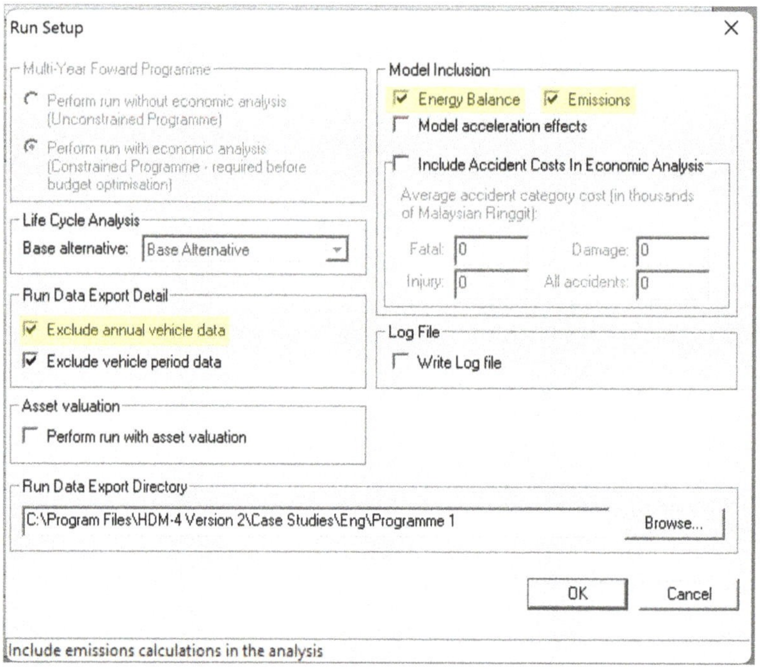

▲ **HDM-4 run setup for vehicle emissions**

The tool kit has been designed to work in Excel 2020 or newer using macros on the Microsoft Windows platform.

When opening the tool kit, you may be given a security warning about active content. If you do not enable the content, the macros in the tool kit will not run and the tool kit will not function correctly.

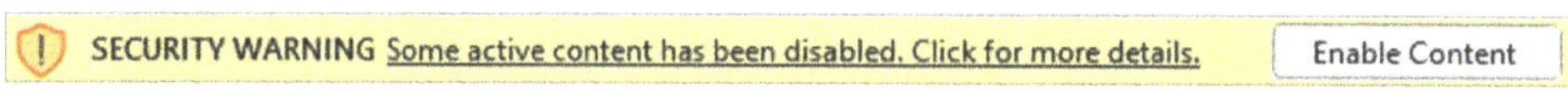

▲ **Enabling macros**

There are six steps in using the tool kit:

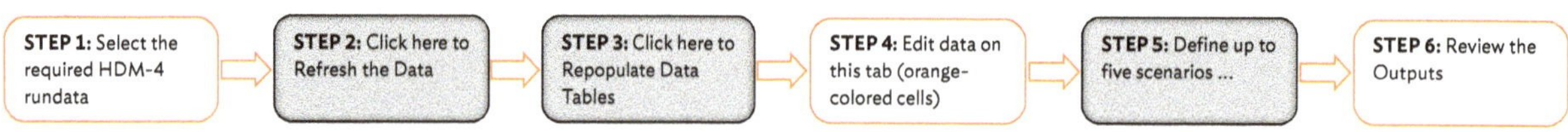

▲ **Six steps in using the tool kit**

Step 1: Select the required HDM-4 run data

Step 2: Refresh the tool kit with the selected HDM-4 run data

Step 3: Repopulate and format the tool kit data tables

Step 4: Enter the input data

Step 5: Define the scenarios

Step 6: Review the outputs

If necessary, repeat steps 4, 5, and 6 to determine the impact of changes in the input data and scenarios.

Each of these steps is described in more detail in the following sections.

B. POPULATING THE TOOL KIT

B.1 Selecting the HDM-4 Run Data

To select the HDM-4 Run Data, go to the ***Input Tables*** sheet and enter the location of the HDM-4 ***RunData.mdb*** file containing the data you wish to analyze.

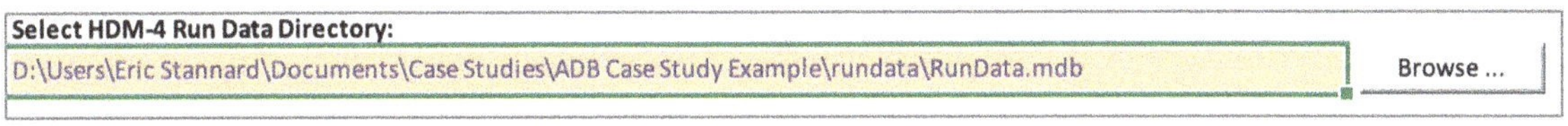

Entering the location of the HDM-4 run data

If you click on the ***Browse*** button, a dialog will pop up allowing you to navigate to the folder; then select the file.

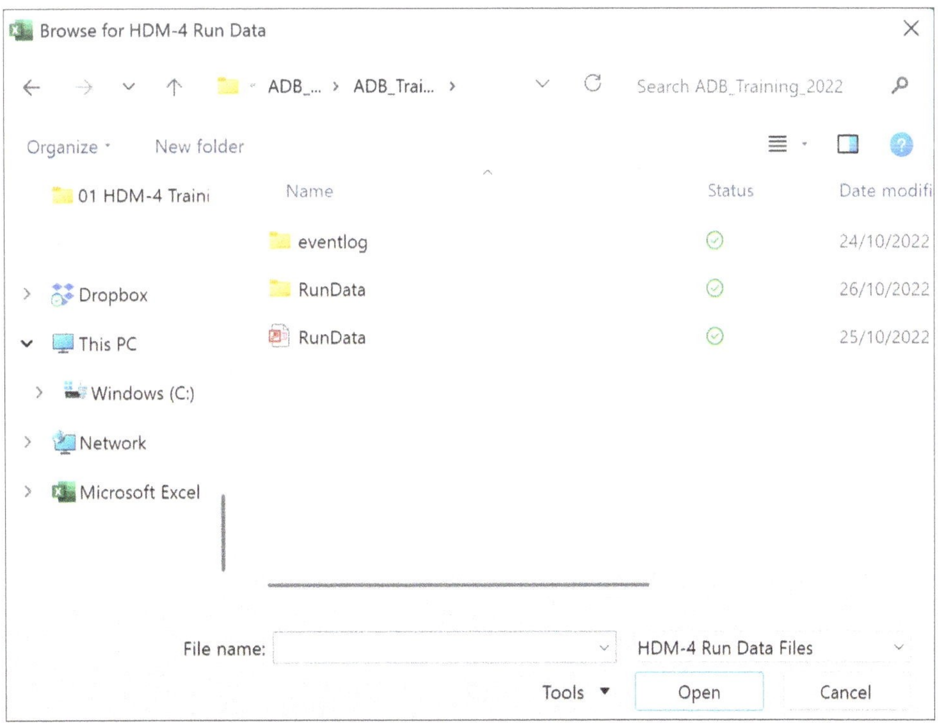

Selecting HDM-4 run data file

If you click on the ***Run Data Summary*** button, you will be presented with details of the analysis contained in the run data so that you can confirm this is the correct file and that the correct models have been selected.

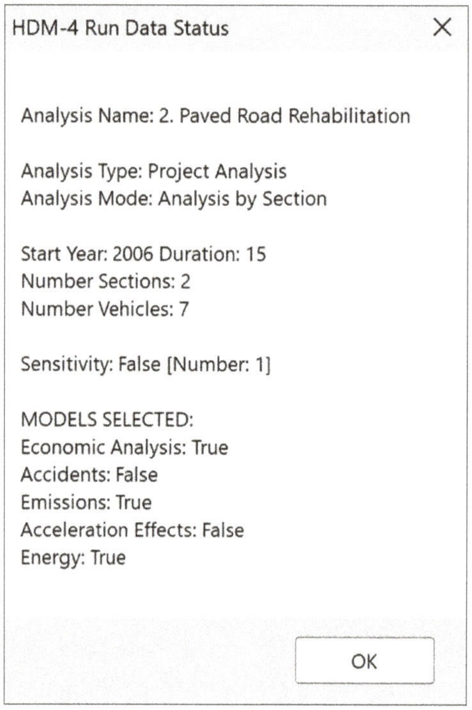

HDM-4 run data status

B.2 Reading the HDM-4 Run Data

The tool kit needs to read the data from the HDM-4 run data file and format it to enable the analysis to take place. Once you have selected the HDM-4 run data, click on the following button to read the data into the tool kit:

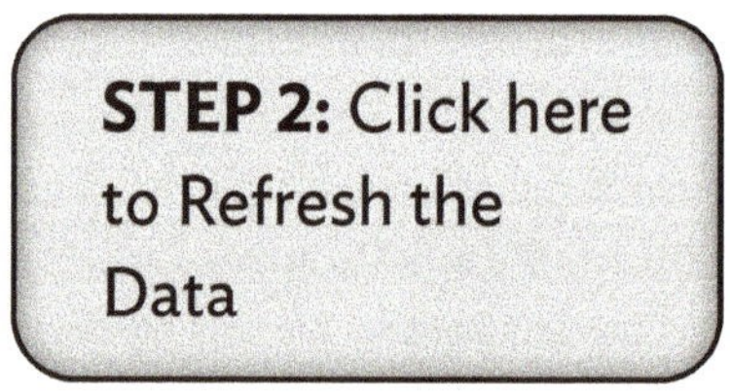

Reading the data from HDM-4 run data file

Reading of the data can take some time depending on the specification of the PC and the size of the run data. While the run data is being read, a spinning curser will be displayed, and the data being read will be displayed in the status bar of the Excel sheet. When the process is complete, ***Ready*** will be displayed in the status bar, and the cursor will return to a pointer.

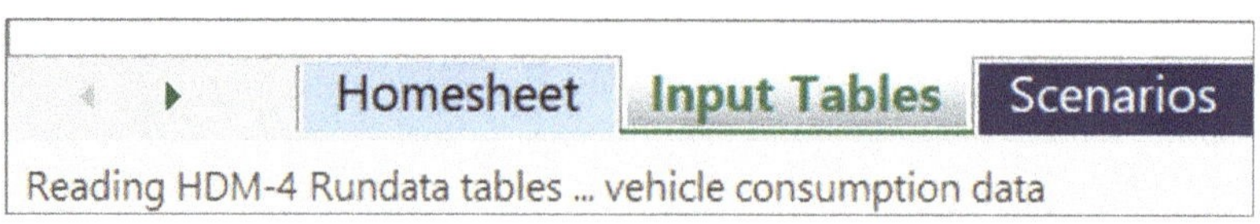

Run data reading status

B.3 Formatting the Tool Kit

To populate the tool kit tables with the data that has been read from HDM-4, and to format the input, analysis, and output sheets, click on the following button:

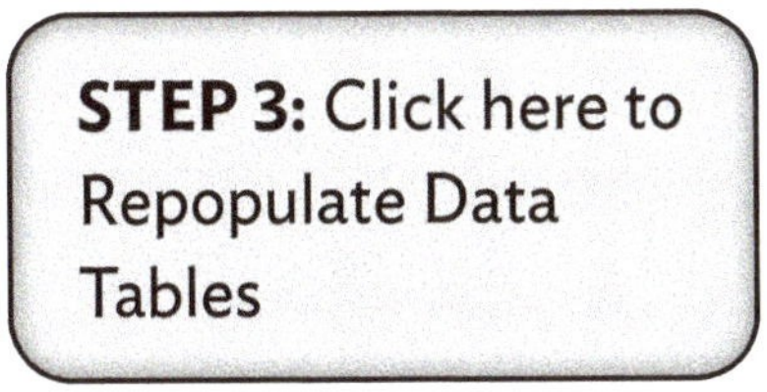

Populating the tool kit with the data read from HDM-4

Once this process has been completed, the tool kit will be fully loaded with the data from the HDM-4 analysis you have prepared.

The input sheet will now display the project title in the header, some details about the analysis, and a list of the work items triggered within the HDM-4 analysis and the vehicles in the vehicle fleet:

Run Data	
Number of Budget Scenarios	5
Number of Years	31
Road length (km)	18,686

HDM-4 run data summary

A warning will be displayed if any of the vehicles within the HDM-4 setup have Energy Used in Production values greater than 0 GJ. This is only an issue if there is an intent to use vehicle transitions (i.e., ICE to EV) for that vehicle type. See Section 2.4.3 and Section 5.4 for details.

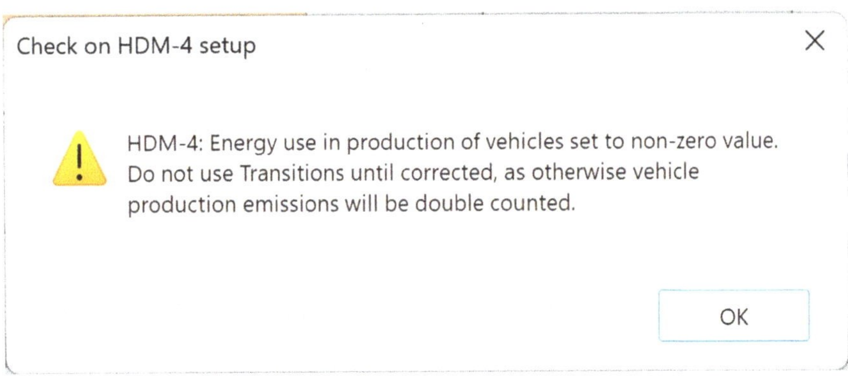

Check on double counting of vehicle production emissions

B.4 Preparing the Input Data: Emissions from Works

The tool kit reads the work items and improvement standards used in the analysis and displays the list on the input sheet.

The number of different work items is shown. The maximum number of work items currently allowed in the tool kit is 100.

The code and name columns show the unique identifiers entered into HDM-4.

You are required to enter the CO_2e emitted by each work activity for the given unit of work.

Note that the units can change depending on the work activity.
It is important that the CO_2e is given for the correct units.

Emissions from Works

Number of Work Items:	26
Maximum Number of Work Items Allowed	100

WORK_CODE	WORK_NAME	kg of CO2 /unit	Unit
	Prep. Edge Repair	35	sq. m
	Prep. Patching	35	sq. m
	Prep. Spot Regravelling	5	cu. m
CS	Crack Sealing	1	sq. m
G2Y1	G2Y4	15	km
G2Y2	G2Y1	15	km
G2Y3	G2Y2	15	km
G2Y4	G2Y3	15	km
grad	grading	3	km
GRAV	Regravelling	20	cu. m
OL	Overlay	35	sq. m
OL	Overlay 4cm	25	sq. m
OL	Overlay 8 cm	50	sq. m
PM	Preventive Overlay	25	sq. m
PO	Preventive Overlay	25	sq. m
POPA	Patching	35	sq. m
R1	Reconstruction C1	260	sq. m
R2	Reconstruction C2	260	sq. m
R3	Reconstruction C3	260	sq. m
Rec	Reconstruction	260	sq. m
REGR	Regravelling2	5	cu. m
REGR2	Gravelbase	7	cu. m
RM	Routine Maintenance	10	km
SDSC	Surface Dressing	15	sq. m
spot2	spot regravel2	20	cu. m
SPOT2	spot2	20	cu. m

Inputs for unit CO_2 emissions from works

In the example above, there are different work activities that have different units. Examples are as follows:

- Edge repair – CO_2e per m^2 (square meter) of work carried out
- Regravelling – CO_2e per m^3 (cubic meter) of work carried out
- Upgrade from unpaved to paved – CO_2e per km of work carried out
- You **must** supply a value for each of the work items listed.

The ***Reference Values*** sheet contains some typical values that can be used for reference purposes.

B.5 Preparing the Input Data: Vehicle Manufacture and Fuel

The tool kit reads the vehicles used in the HDM-4 analysis and displays the list on the input sheet.

The number of different vehicles is shown. The maximum number of vehicles allowed is 50.
The ***Well-Tank Adjustment Factor*** attribute is also shown, and the default value should be reviewed and changed if necessary.

HDM-4 computes the tailpipe carbon emissions for the vehicles but does not compute the carbon cost of fuel production and distribution. The well-to-tank emission multiplication factor applied to the tailpipe emissions include the GHG emissions released into the atmosphere from production, processing, and delivery of fuel. The recommended default value for the well-to-tank adjustment factor is 1.3.

Vehicle Manufacture & Fuel

Number of Vehicles:	7
Well-Tank Adjustment Factor	1.3

Vehicle ID	Vehicle Name	kg of CO_2e /vehicle	ICE/EV/NMT	HDM-4 CO_2 Included (ICE)	HDM-4 CO_2 Included (EV)	Vehicle Category
1	F- Medium Truck 3-Axles 8-14t	24,000	ICE	100%	100%	Medium Truck
2	E- Small 2-axle Truck 2-8t	17,000	ICE	100%	100%	Light Truck
3	H- Articulated Vehicle >3 axles >14t	33,000	ICE	100%	100%	Heavy Truck
4	D- Van up to 2t	11,800	ICE	100%	100%	SUV/Utility
5	C- Bus	16,200	ICE	100%	100%	Bus
6	B- Minibus	11,800	ICE	100%	100%	SUV/Utility
7	A- Personal Car	9,200	ICE	100%	100%	Car

▲ **Inputs for vehicle manufacture and well-to-tank factor**

The vehicle ID (number) and vehicle name columns show the details of each vehicle as entered in HDM-4.

You are required to enter the following information:

- CO_2e emitted in kilograms during the manufacture of each vehicle type.
- The vehicle type from a list is displayed when you highlight the cell. Choose between Internal Combustion Engine (ICE), Electric Vehicle (EV), Hybrid, or Nonmotorized Transport (NMT). The selection will ensure the tool kit uses the correct data from HDM-4 to calculate the carbon impact of each vehicle.

  The default operation is for the tool kit to use the emission values computed by HDM-4 for ICE vehicles; for EVs, the energy values are converted to a carbon equivalent.

  This setting should be used in conjunction with the "HDM-4 CO_2 Included" column.
- "HDM-4 CO_2 Included" percentage for each vehicle. This value allows you to adjust the percentage of tailpipe or EV CO_2 calculated by HDM-4 to be included in the tool kit calculations.

  For example, for "HDM-4 CO_2 included (ICE)":

  Tool Kit Vehicle Emissions = HDM-4 Vehicle Emissions * (Percent Included/100)

  Therefore: 0% – no emissions included from HDM-4 (EV)

  100% – all emissions included from HDM-4 (ICE)

  70% – % tailpipe emissions computed by HDM-4 are included (Hybrid)
- Vehicle Category. Assign each vehicle to one of the predefined vehicle categories. This categorization is used in a more advanced analysis performed by the tool kit to determine the impact of shifts of the vehicle to low emission types (see Section 7 on scenario analysis).

You **must** supply a value for each of the vehicles listed.

The ***Reference Values*** sheet contains some typical values that can be used for reference purposes.

B.6 Preparing the Input Data: Economic Parameters

The output currency and discount rate used in the HDM-4 analysis are displayed under the Economic Parameters section within the Input Tables sheet.

The tool kit requires CO_2 cost per ton to be entered, which is used in the tool kit to calculate the total carbon cost of the scenarios defined. This cost should be entered in the same currency as the HDM-4 output currency.

Economic Parameters	
Discount Rate	10%
Currency	US Dollar
Standard Conversion Factor (SCF)	0.85
CO_2 Costs/ton	42
Currency	US Dollar

▲ **Inputs for economic parameters**

B.7 Preparing the Input Data: Ancillary Assets and Activities

The tool kit allows you to specify the carbon and economic cost of maintaining ancillary assets or performing activities that are not directly included in an HDM-4 analysis.

Assets and activities are user defined. For each asset and activity, specify the following:

- Asset and/or activity name – identify a unique name and the work item it represents.
- Unit – this represents the unit of work performed. For example:
  - EV charging infrastructure – No. – in the scenario worksheet, you would enter the number of charging infrastructure built and each of these would have a cost of $60,000 and emit 100,000 kg of CO_2e
  - NMT lane – km – in the scenario worksheet, you would enter the length in km of NMT lanes that are to be built/maintained, and each km of work would have a cost of $100,000 and emit 1,000 kg of CO_2e
  - Cost per unit – the cost in HDM-4 output currency of one unit of activity
  - CO_2e emissions in kg per unit of the activity

If you do not intend to use any additional assets or activities in the tool kit analysis, these can be left blank.

Ancillary Assets & Activities (Not included in HDM-4)

No.	Asset /Activity	Unit	Costs/Unit(US Dollar)	Kg CO_2e/Unit
1	EV Charging Station (Level 3)	No.	60,000,000	500
2	Bridge (Concrete)	No.	123,456	100,000
3	Tunnel	No.	6,000,000	100,000
4	NMT Lane	km	100,000	1,000
5	Bridge Maintenance	No.	30,000	2,000
6	Tunnel Maintenance	No.	20,000	2,000
7	undefined			
8	undefined			
9	undefined			
10	undefined			

▲ **Inputs for ancillary assets and other activities**

C. SETTING UP SCENARIOS

C.1 Introduction

On the *Scenarios* worksheet, you can define five scenarios to analyze within the tool kit.

By default, the first scenario is assumed to be the base scenario for the purpose of economic evaluation.

The process of setting up all scenarios is the same for all five and is described below.

Grid Factor	Select Country	Grid Factor kgCO2e/kWh	Year	Source
	Kyrgyz Republic	0.28	2015	IEA (2017) for GHG; IEA for electricity production minus losses

☑ Show Transition to LZEV Details

						1	2	3	4	5	6	7	8
No.	Scenario Name	HDM Investment Alternative	Criteria	Unit	Total	2023	2024	2025	2026	2027	2028	2029	2030
1	Business As Usual	Budget (100%)	**Grid Factor**	$kgCO_2e/kWh$		0.28	0.28	0.28	0.28	0.28	0.28	0.28	0.28
			Carbon Reduction Scheme	CO_2e (kg)		-	-	-	-	-	-	-	-
			Transition to LZEV	%		-	-	-	-	-	-	-	-
			Motorcycle										
			Car										
			SUV/Utility										
			Bus										
			Light Truck										
			Medium Truck										
			Heavy Truck										
			Rickshaw										
			Pedestrian										
			Bicycle										
			Other Asset / Activities:										
			EV charging infrastructure	No.			10.00	20.00	25.00	50.00	80.00	80.00	100.00
			NMT Lane	km									

▲ **Inputs for defining analysis scenarios**

When entering the data to set up the scenarios, the analysis results are updated each time you make a change. You can therefore view the impact on the results as you make incremental changes to the scenarios.

C.2 Scenario Name and HDM-4 Investment Alternative

Each scenario should be given a unique name to identify it in the output reports.

The format of the name should be chosen such that it conveys the scenario that it represents. This makes interpreting the results easier when they are viewed and helps other users better understand the data.

No.	Scenario	HDM Investment Alternative
1	Business as Usual	Budget (100%)

◀ **Naming scenarios**

For each scenario, select the HDM-4 Investment Alternative to be used for the core data (road work program and vehicle fleet emissions).

To select the investment alternative, click the appropriate cell, then click the down arrow symbol to display the list of alternatives imported with the HDM-4 data.

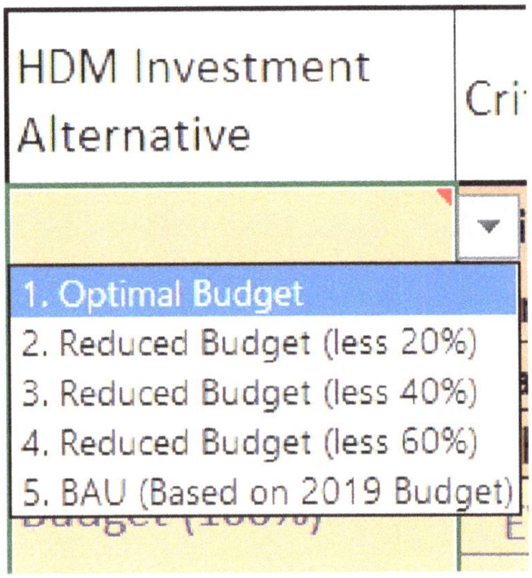

Selecting investment alternative for a scenario

In the case of HDM-4 Program or Strategy Analysis being imported, the budget scenarios defined in the analysis will be displayed.

In the case of a Project Analysis being imported, the project alternatives will be displayed. If you have selected a Project Analysis, you should ensure that the Base Alternative is selected for the first scenario.

C.3 Grid Factor

The impact of EVs and their electricity consumption is related to the actual grid emissions, reflecting the emissions caused by EVs in the same manner as other electricity-consuming devices in the same country. The grid factor defines the amount of CO_2e emissions produced per unit of electricity generated. The grid emission factor or grid factor can be derived based on the total electricity produced minus losses and the total GHG emissions of the electricity production sector.

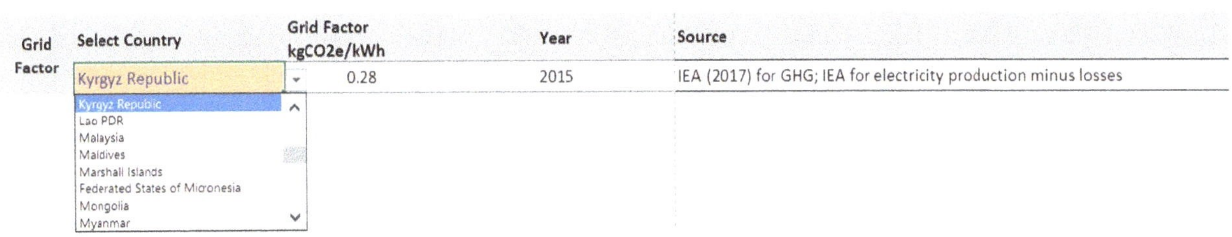

Viewing reference grid factors

By clicking the ***Select Country*** input, a country can be selected from the list. A default grid factor value will be displayed together with the details of when this value was derived. This value can be used as a reference when entering the data for the grid factor within each scenario.

For each year of each scenario, enter the grid factor.

			1	2	3	4	5
Criteria	Unit	Total	2023	2024	2025	2026	2027
Grid Factor	$kgCO_2e/kWh$		0.28	0.28	0.28	0.28	0.28

Entering grid factors for each scenario

Hint: The grid factor does not have to be the same for each year of the analysis. By varying the grid factor through the years, you can represent changes in the emissions produced using cleaner energy. Definition of future grid factors for a country should consider the likely pace of greening the grid.

C.4 Carbon Reduction Scheme

Alongside road schemes, carbon reduction schemes are often introduced to offset the impact of the maintenance, improvement, or construction of roads. The tool kit allows you to enter the CO_2 reduction (saving) of the scheme to be considered as part of the scenario.

The carbon reduction values should be entered for each year of the analysis.

			1	2	3	4	5
Criteria	Unit	Total	2023	2024	2025	2026	2027
Grid Factor	$kgCO_2e/kWh$		0.28	0.28	0.28	0.28	0.28
Carbon Reduction Scheme	CO_2e (kg)		1,000	1,200	1,400	1,600	1,800

Entering inputs for carbon reduction scheme

The tool kit does not include a calculator to determine the values of carbon reduction schemes, thus the data should be calculated external to the tool kit. Various calculators are available externally. For example, the Asian Development Bank has produced a Sequestration Calculator that allows you to estimate the carbon reduction from different tree planting schemes.[1]

1 Link to CO_2-Calculator?

C.5 Transition to Low- or Zero-Emission Vehicle

The tool kit allows you to model the transition of the existing vehicles modeled in HDM-4 to low- or zero-emission vehicles to determine the impact on the overall life cycle carbon emissions for the scenario.

If you are not including this feature within your scenarios, you can untick the ***Show Transition to LZEV Details*** box on the Scenarios tab to reduce the complexity of the inputs. The detailed data will be hidden but will not be deleted. With the box unticked, the outputs of the analysis will not include the results of the transition analysis.

If you wish to include transition to LZEVs in the outputs of the analysis and to see the LZEV transition details, ensure that the box is ticked.

☑ Show Transition to LZEV Details

Including transition to LZEVs in scenario definition and in outputs

If you are using transitions, then on the Transitions tab, set up the rate at which vehicles will transition from one type (ICE) to another (EV).

Transition Scenarios

No.	Name	Profile	Edit
1	No Transition		Edit / View
2	Light Vehicles to EV		Edit / View
3	Medium		Edit / View
4	Fast		Edit / View
5	For Testing		Edit / View
6	No Transition		Edit / View
7	Slow Transition		Edit / View
8	Fast		Edit / View
9	Very Fast		Edit / View
10	Undefined		Edit / View

Setting up transitions – summary

						1	2	3	4	5	6	7	8
No.	**Transition Scenario Name**	**Description**	**Criteria**	**Unit**	**Total**	**2022**	**2023**	**2024**	**2025**	**2026**	**2027**	**2028**	**2029**
1	No Transition	No transition of vehicle types	**Transition to LZEV**	%		-	-	-	-	-	-	-	-
			Motorcycle										
			Car										
			SUV/Utility										
			Bus										
			Light Truck										
			Medium Truck										
			Heavy Truck										
			Rickshaw										
			Pedestrian										
			Bicycle										
2	Light Vehicles to EV	Light vehicles transition to EVs by 2040	**Transition to LZEV**	%		1.50%	1.80%	2.10%	2.40%	2.70%	3.00%	3.60%	4.50%
			Motorcycle			5.00%	6.00%	7.00%	8.00%	9.00%	10.00%	12.00%	15.00%
			Car			5.00%	6.00%	7.00%	8.00%	9.00%	10.00%	12.00%	15.00%
			SUV/Utility			5.00%	6.00%	7.00%	8.00%	9.00%	10.00%	12.00%	15.00%
			Bus			0.00%	0.00%	0.00%	0.00%	0.00%	0.00%	0.00%	0.00%
			Light Truck			0.00%	0.00%	0.00%	0.00%	0.00%	0.00%	0.00%	0.00%
			Medium Truck			0.00%	0.00%	0.00%	0.00%	0.00%	0.00%	0.00%	0.00%
			Heavy Truck			0.00%	0.00%	0.00%	0.00%	0.00%	0.00%	0.00%	0.00%
			Rickshaw			0.00%	0.00%	0.00%	0.00%	0.00%	0.00%	0.00%	0.00%
			Pedestrian			0.00%	0.00%	0.00%	0.00%	0.00%	0.00%	0.00%	0.00%
			Bicycle			0.00%	0.00%	0.00%	0.00%	0.00%	0.00%	0.00%	0.00%

▲ **Setting up transitions – details**

With the details shown, the tool kit will display the 10 default vehicle categories. These are fixed in the tool kit and cannot be changed.

For each vehicle, for each year, enter the percentage of that vehicle category that you wish to transition to an LZEV equivalent vehicle.

			1	2	3	4
Criteria	Unit	Total	2023	2024	2025	2026
Transition to LZEV	%		-	-	-	-
Motorcycle						
Car						
SUV/Utility						
Bus						
Light Truck						
Medium Truck						
Heavy Truck						
Rickshaw						
Pedestrian						
Bicycle						

▲ **Defining the percentage of vehicles transitioning to LZEVs**

For instance, if you wish to investigate the scenario where 5% of vehicles classified as ***Car*** move to a low- or zero-emission vehicle in the first year of the analysis, and an additional 5% each year thereafter, with 2% of ***Bus*** types doing the same, the inputs would be as follows:

			1	2	3	4
Criteria	Unit	Total	2023	2024	2025	2026
Transition to LZEV	%		3.50%	7.00%	10.50%	14.00%
Motorcycle						
Car			5.00%	10.00%	15.00%	20.00%
SUV/Utility						
Bus			2.00%	4.00%	6.00%	8.00%
Light Truck						
Medium Truck						
Heavy Truck						
Rickshaw						
Pedestrian						
Bicycle						

Illustration of definition of transition to LZEVs

Note that the *Transition to LZEV* row contains the average of all vehicle transitions.

C.6 Other Assets and Activities

To include the impact of other assets and activities in a scenario, you can select up to five different categories as defined on the *Input Data* sheet.

After selecting the asset and/or activity, define the units of activity performed each year. The unit will depend on the selection made.

			1	2	3	4
Criteria	Unit	Total	2023	2024	2025	2026
Other Asset / Activities:						
EV charging infrastructure	No.			10.00	20.00	25.00
NMT Lane	km		100.00	120.00	140.00	100.00

EV charging infrastructure
NMT Lane
Tunnel
undefined
undefined
undefined
undefined
undefined

Including the impact of other assets and activities

To select an asset and/or activity, click within the cell, and a down arrow will be displayed. Click the down arrow to display the list. Select an item from the list.

If you wish to delete the asset, highlight the cell and click ***delete*** to remove the selection and repeat for all the years, so no data is defined.

Note: If you change the assets and/or activities in the ***Input Data*** sheet, revisit the ***Scenarios*** sheet to ensure the data and selections are still valid.

D. ANALYZING AND INTERPRETING OUTPUTS

D.1 Introduction

The detailed results of the analysis for the scenarios are available in the following sheets (color-coded yellow ■):

- Analysis – HDM4 Works Emissions
- Analysis – Vehicle Emissions
- Analysis – EV Vehicle Emissions
- Analysis – Vehicle Manufacturing Emissions
- Analysis – External Calculations
- Analysis – External Economics
- Analysis – Economics
- Analysis – Transition Emissions

The output reports are contained in the following sheets (color-coded blue ■):

- Output – Total Emission Graphs
- Output – Annual Emissions
- Annual Emission Graphs
- Output – Annual Vehicle Data
- Output – Vehicle Speeds
- Output – Economics

Each of the analysis/output sheets is discussed below.

D.2 Analysis: HDM-4 Works Emissions

This sheet contains the calculations for the CO_2e emissions for the program of works for each scenario.

CO_2e from Physical Works

Scenario: Business as Usual

No.	Alternative	Intervention	Treatment Code	CO_2/Unit (kg)	Works Unit	Annual emissions for treatment for each year of analysis for scenario, CO_2e (tons) Total	1 2023
1	5. BAU (Based on	Prep. Edge Repair	0	35	sq. m	1,271	397
2	5. BAU (Based on	Prep. Patching	0	35	sq. m	99,420	6,725
3	5. BAU (Based on	Prep. Spot Regravelling	0	5	cu. m	12	0
4	5. BAU (Based on	AMGB Recon/Heavy Rehab	AMREC	50	sq. m	–	0
5	5. BAU (Based on	Edge Repair	EDGE	35	sq. m	17,836	7
6	5. BAU (Based on	(I2) Capacity Improvement	KGZI	130	sq. m	800,449	181,528
7	5. BAU (Based on	(I1) Upgrade Unpaved to Paved @ 300 vpd	KUUP	1,000,000	km	458,400	0
8	5. BAU (Based on	Overlay 50 mm	OVL5	70	sq. m	1,137,356	0
9	5. BAU (Based on	Patching	PATCH	35	sq. m	2,058	0
10	5. BAU (Based on	Pothole Patching	POTH	35	sq. m	131,878	32
11	5. BAU (Based on	STGB Recon/HeavyRehab	RECON	45	sq. m	–	0
12	5. BAU (Based on	Regravelling	REG	20	cu. m	872	0
13	5. BAU (Based on	Routine Miscellaneous	ROUT_U	5	km	1,592	53
14	5. BAU (Based on	Routine Miscellaneous	ROUTB	5	km	1,592	53
15	5. BAU (Based on	Routine Miscellaneous	ROUTM	5	km	1,592	53
16	5. BAU (Based on	Resealling	RSL13	45	sq. m	31,188	0
					TOTAL	2,685,516	188,848

▲ **Analysis worksheet for works emissions**

The work activities and work quantities are imported from HDM-4 and the CO_2e is calculated based on the CO_2 emission value entered in the ***Input Data*** sheet for each work item.

The total CO_2e for each work item is given in each row in the ***Total*** column.

The total works CO_2e for the scenario, and for each year, is given in the ***TOTAL*** row.

By scrolling down the sheet, you can see the results for each of the scenarios defined. To quickly jump to a scenario, use the drop-down list to automatically navigate to the start of the results of the scenarios:

By looking at the results in this sheet, you can determine the individual contributions to the total works CO_2e, allowing you to highlight the work items that have the most impact on the total emission values.

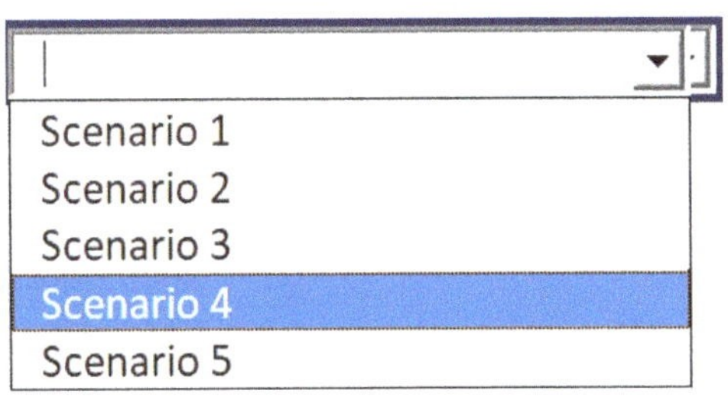

▲ **Navigating analysis worksheets**

D.3 Analysis: Internal Combustion Engine Vehicle Emissions

This sheet contains the calculations for the CO_2e emissions for the vehicle types for each scenario based on the fuel used for the journeys.

CO_2e from Vehicles

Business as Usual				Manufacture			Annual CO_2
							1
No.	Alternative	ICE/EV/NMT	ICE/EV/NMT	CO_2/Unit (ton)	% CO_2	Total	2023
1	5. BAU (Based on	F – Medium Truck 3-Axles 8-14t	ICE	24	100%	11,391,456	184,936
2	5. BAU (Based on	E – Small 2-axle Truck 2-8t	ICE	17	100%	7,705,593	130,170
3	5. BAU (Based on	H – Articulated Vehicle >3 axles >14t	ICE	33	100%	40,982,339	686,948
4	5. BAU (Based on	D – Van up to 2t	ICE	11.8	100%	25,334,625	409,913
5	5. BAU (Based on	C – Bus	ICE	16.2	100%	56,318	1,006
6	5. BAU (Based on	B – Minibus	ICE	11.8	100%	12,814,900	208,233
7	5. BAU (Based on	A – Personal Car	EV	9.2	100%	212,499,761	2,821,043
					TOTAL	310,784,991	4,442,250

▲ **Analysis worksheet for vehicle emissions**

The tailpipe emission quantities are imported from HDM-4 and the CO_2e is calculated based on the well-to-tank (WTT) adjustment factor, and the percentage of HDM-4 emissions to include for a vehicle type in the ***Input Data*** sheet for each vehicle type and scenario.

$$ICE\_CO_2e_{vys} = WT \times TP\_CO_2e_{vys} \times pCO_2e_{vys}/1000$$

Where:

$ICE\_CO_2e_{vys}$ = fuel emissions expressed in tons

v = vehicle type

y = year

s = scenario

WT = the adjustment factor to account for the well-to-tank (≥1.0)

$TP\_CO_2e_{vys}$ = the predicted HDM-4 tailpipe emissions for ICE vehicles

pCO_2e_v = the percentage of HDM-4 emissions to include for a vehicle type

The total CO_2e for each vehicle type is given in each row in the ***Total*** column.

The total works CO_2e for the scenario, and each year is given in the ***TOTAL*** row.

By scrolling down the sheet, you can see the results for each of the scenarios defined. To quickly jump to a scenario, you can use the drop-down list located at the top of the worksheet to automatically navigate to the start of the results of the scenarios.

By looking at the results in this sheet, you can determine the individual contributions to the total vehicle CO_2e from fuel usage allowing you to highlight the vehicles that have the highest impact on the total emission values.

D.4 Analysis: Electric Vehicle Emissions

This sheet contains the calculations for the CO_2e emissions for the electric vehicle types for each scenario based on the total energy used.

TotalOffset

CO_2 from Vehicles ☐ Vehicle Manufacture CO_2 YearOffset

☑ Electric Vehicle CO_2 **Annual CO_2**

Business as Usual 1

No.	Alternative	Vehicle Name	ICE/EV/NMT	Manufacture CO_2/Unit (ton)	% CO_2	Total	2023
1	5. BAU (Based on	F – Medium Truck 3-Axles 8-14t	ICE	24	100%	0	0
2	5. BAU (Based on	E – Small 2-axle Truck 2-8t	ICE	17	100%	0	0
3	5. BAU (Based on	H – Articulated Vehicle >3 axles >14t	ICE	33	100%	0	0
4	5. BAU (Based on	D – Van up to 2t	ICE	11.8	100%	0	0
5	5. BAU (Based on	C – Bus	ICE	16.2	100%	0	0
6	5. BAU (Based on	B – Minibus	ICE	11.8	100%	0	0
7	5. BAU (Based on	A – Personal Car	EV	9.2	100%	243,855	6,994
					SUM	243,855	6,994

▲ **Analysis worksheet for electric vehicle emissions**

The emission quantities are calculated from the vehicle's energy usage data imported from HDM-4. The CO_2e is calculated by converting the energy usage from HDM-4 (in joules) to kilowatt-hours and the grid factor for the scenario/year.

$$EV\_CO_2e_{vys} = MJ_{vys} \times ECF \times Grid\_Factor \times pCO_2e_{vys}/1000$$

Where MJ_{vys} is the total number of MJ of energy usage for EVs from HDM-4.

MJ_{vy} is available by querying the output HDM-4 database and by looking for parameter ENGFACT_12 within the Annual Vehicle Table.

ECF is the factor for converting energy usage from HDM-4 (in joules) to kilowatt-hours

$Grid\_Factor$ defines the CO_2e emissions produced by generating 1 kWh of electricity.

The total CO_2e for each vehicle type item is given in each row in the ***Total*** column.

The total vehicle type CO_2e for the scenario, and each year is given in the ***TOTAL*** row.

By scrolling down the sheet, you can see the results for each of the scenarios defined. To quickly jump to a scenario, use the drop-down list located at the top of the worksheet to automatically navigate to the start of the results of the scenarios.

By looking at the results in this sheet, you can determine the individual contributions to the total electric vehicle CO_2e from energy usage, allowing you to highlight the vehicles that have the highest impact on the total emission values.

D.5 Analysis: Low- or Zero-Emission Vehicle Transition Emissions

This sheet contains the calculations for the CO_2e emissions for transitioning ICE vehicles to EVs for each vehicle type scenario based on the annual transition percentages specified by the user in the ***Scenarios*** worksheet.

						ScenarioOffset TotalOffset YearOffset **Annual CO_2e emissions for each year of analysis for selected budget (tons)**	
CO_2e from Vehicles							
Business as Usual				Manufacture			1
No.	Alternative	Vehicle Name	ICE/EV/NMT	CO_2e/Unit (ton)	Energy Factor	Total	2023
1	5. BAU (Based on	F – Medium Truck 3-Axles 8-14t	EV	24	100%	11,391,456	184,936
2	5. BAU (Based on	E – Small 2-axle Truck 2-8t	EV	17	100%	7,705,593	130,170
3	5. BAU (Based on	H – Articulated Vehicle >3 axles >14t	EV	33	100%	40,982,339	686,948
4	5. BAU (Based on	D – Van up to 2t	EV	11.8	100%	25,334,625	409,913
5	5. BAU (Based on	C – Bus	EV	16.2	100%	34,686	986
6	5. BAU (Based on	B – Minibus	EV	11.8	100%	12,814,900	208,233
7	5. BAU (Based on	A – Personal Car	EV	9.2	100%	38,302,552	2,680,341
					SUM	136,566,151	4,301,527

▲ **Calculations for the CO_2e emissions for transitioning vehicles**

Emission quantities following transition to LZEV is calculated by applying the grid factor specified by the user and the proportion of each ICE vehicle transitioning to EVs specified in the ***Scenario*** worksheet to the Energy Usage of EVs as follows:

$$EV\_CO_2e_{vys} = MJ_{vys} \times ECF \times Grid_{Factor} \times pCO_2e_{vys} \times pTLZEV_{vys}/1000$$

Where: $pTLZEV_{vys}$ = percentage of ICE vehicles transitioning to EVs for each vehicle type in each year and for each scenario.

The remaining tailpipe vehicle emissions following the transition of a proportion of vehicles to EVs is calculated.

$$ICE\_CO_2e_{vys} = WT \times TP_{CO_2e_{vys}} \times pCO_2e_{vys} \times (1 - pTLZEV_{vys})/1000$$

D.6 Analysis: Vehicle Manufacturing Emissions

This sheet contains the calculations for the CO_2e emissions associated with the manufacture of the vehicle types for each scenario based on the average life of a vehicle and distance driven as defined in HDM-4.

The estimate of the vehicles to be replaced during the life cycle analysis are imported from HDM-4 and the CO_2e is calculated based on the carbon emission rates entered in the ***Input Data*** sheet for the manufacture of each vehicle type.

The total CO_2e for each vehicle is given in each row in the ***Total*** column.

The total vehicle CO_2e for the scenario, and each year is given in the ***TOTAL*** row.

CO_2e from Vehicles				☑ Vehicle Manufacture CO_2 ☐ Electric Vehicle CO_2		TotalOffset 5 YearOffset 5 Annual CO_2e emissions for each year of analysis for selected budget (tons)	
Business as Usual				Manufacture		1	2
No.	Alternative	Vehicle Name	ICE/EV/NMT	CO_2/Unit (ton)	Total	2023	2024
1	5. BAU (Based on	F – Medium Truck 3-Axles 8-14t	ICE	24	257,291	3,934	4,094
2	5. BAU (Based on	E – Small 2-axle Truck 2-8t	ICE	17	632,373	9,675	10,069
3	5. BAU (Based on	H – Articulated Vehicle >3 axles >14t	ICE	33	1,079,879	16,725	17,470
4	5. BAU (Based on	D – Van up to 2t	ICE	11.8	3,305,959	50,122	52,394
5	5. BAU (Based on	C – Bus	ICE	16.2	2,334	36	38
6	5. BAU (Based on	B – Minibus	ICE	11.8	1,602,156	24,292	25,396
7	5. BAU (Based on	A – Personal Car	EV	9.2	50,533,691	639,124	675,577
				TOTAL	57,413,682	743,909	785,038

Analysis worksheet for vehicle manufacture emissions

By scrolling down the sheet, you can see the results for each of the scenarios defined. To quickly jump to a scenario, use the drop-down list located at the top of the worksheet to automatically navigate to the start of the results of the scenarios.

By looking at the results in this sheet, you can determine the individual contributions to the total vehicle CO_2e from vehicle manufacture, allowing you to highlight the vehicles that have the highest impact on the total emission values.

D.7 Analysis: External Calculations

This sheet contains the calculations for the CO_2e emissions associated with the assets and activities defined by the user in the ***Scenarios*** sheet. Emission factors for each ancillary asset and/or activity provided by the user in the ***Input*** sheet are applied to quantities entered in the ***Scenarios*** worksheet to calculate emissions associated with ancillary assets and other activities.

CO_2e from External Calculations							Annual CO_2e emissions (tons)	
Business as Usual							1	2
No.	Alternative	Intervention	Treatment Code	CO_2/Unit (kg)	Works Unit	Total	2023	2024
1	5. BAU (Based on 2019 Budget)	EV charging infrastructure		5,000,000	No.	312,500	0	5,000
2	5. BAU (Based on 2019 Budget)	NMT Lane		1,000	km	470	100	120
3	5. BAU (Based on 2019 Budget)					0		
4	5. BAU (Based on 2019 Budget)					0		
5	5. BAU (Based on 2019 Budget)					0		
					TOTAL	312,970	100	5,120

Analysis of emissions for ancillary assets and activities

The total CO_2e for each asset and/or activity is given in each row in the ***Total*** column.

The total asset and/or activity CO_2e for the scenario, and each year is given in the ***TOTAL*** row.

By scrolling down the sheet, you can see the results for each of the scenarios defined. To quickly jump to a scenario, you can use the drop-down list located at the top of the worksheet to automatically navigate to the start of the results of the scenarios.

By looking at the results in this sheet, you can determine the individual contributions to the total CO_2e from external activities, allowing you to highlight the activities that have the highest impact on the total emission values.

D.8 Analysis: External Economics

This sheet contains the calculations for the undiscounted economic cost with the assets and activities defined by the user in the ***Scenarios*** sheet.

CO_2e from External Calculations

Business as Usual

							Annual CO_2e emissions (tons)	
							1	2
No.	Alternative	Intervention	Treatment Code	CO/Unit (US Dollars)	Works Unit	Total	2023	2024
1	5. BAU (Based on 2019 Budget)	EV charging infrastructure		60,000	No.	37,500,000	0	600,000
2	5. BAU (Based on 2019 Budget)	NMT Lane		100,000	km	47,000,000	10,000,000	12,000,000
3	5. BAU (Based on 2019 Budget)					0		
4	5. BAU (Based on 2019 Budget)					0		
5	5. BAU (Based on 2019 Budget)					0		
					TOTAL	84,500,000	10,000,000	12,600,000

▲ **Undiscounted economic costs for ancillary assets and activities**

Undiscounted economic costs for ancillary assets and activities are derived using unit costs for asset and/or activity specified by the user of the tool kit in the input sheet, the Standard Conversion Factor (SCF) provided in the input sheet, and the quantities provided for each scenario in the ***Scenarios*** worksheet.

$$E\_costs_{ays} = A\_quant_{ays} \times U\_costs_{a} \times SCF$$

Where E_costs_{ays} Undiscounted economic costs calculated for each asset or activity per year and for each scenario

A_quant_{ays} Quantity of asset or activity for each year and scenario

U_costs_{a} Unit costs for each asset or activity

SCF Standard Conversion Factor used to convert financial costs into economic costs.

The total economic cost for each asset and/or activity is given in each row in the ***Total*** column.
The total asset and/or activity economic cost for the scenario, and for each year is given in the ***TOTAL*** row.

D.9 Analysis: Economics

This sheet contains the calculations for the discounted economic cost for the five scenarios defined in the tool kit.

The road agency and road user costs are imported from HDM-4 and displayed on the sheet with an extra column, ***Additional Costs***. This column contains the costs for the assets and activities defined within the tool kit.

Discounted Costs (in millions of US Dollar)

Scenario 0: BAU

Budget (100%)

Analysis Year	Capital Costs	Recurrent Costs	Special Costs	Total RAC	Normal Traffic Motorised VOC	Normal Traffic Motorised TTC	Normal Traffic Non-Motorised TOC	Generated Motorised VOC	Generated Traffic Motorised TTC	Generated Traffic Non-Motorised TOC	Accident Costs	Total RUC	Exogenous Costs	Additional Costs	Total Transport Costs
2022	891.522	12.846	0.000	904.368	2246.471	583.655	0.000	0.000	0.000	0.000	0.000	2830.126	0.000	0.123	3734.618
2023	73.922	19.014	0.000	92.937	1872.854	457.136	0.000	0.000	0.000	0.000	0.000	2329.990	0.000	0.224	2423.151
2024	4.541	11.629	0.000	16.169	1764.540	424.104	0.000	0.000	0.000	0.000	0.000	2188.644	0.000	0.306	2205.120
2025	15.262	11.754	0.000	27.016	1691.371	414.288	0.000	0.000	0.000	0.000	0.000	2105.659	0.000	0.371	2133.046
2026	0.000	10.483	0.000	10.483	1611.803	397.105	0.000	0.000	0.000	0.000	0.000	2008.907	0.000	0.422	2019.812
2027	0.324	10.271	0.000	10.595	1540.562	383.684	0.000	0.000	0.000	0.000	0.000	1924.246	0.000	0.460	1935.300
2028	7.133	9.340	0.000	16.472	1471.933	370.158	0.000	0.000	0.000	0.000	0.000	1842.092	0.000	0.488	1859.052
2029	5.251	8.805	0.000	14.057	1406.432	357.584	0.000	0.000	0.000	0.000	0.000	1764.016	0.000	0.507	1778.580
2030	10.308	7.910	0.000	18.219	1343.591	346.577	0.000	0.000	0.000	0.000	0.000	1690.168	0.000	0.000	1708.387
2031	23.934	6.754	0.000	30.688	1288.366	338.222	0.000	0.000	0.000	0.000	0.000	1626.588	0.000	0.000	1657.275
2032	9.225	7.022	0.000	16.247	1233.727	328.709	0.000	0.000	0.000	0.000	0.000	1562.436	0.000	0.000	1578.683
2033	48.433	6.662	0.000	55.095	1185.593	320.640	0.000	0.000	0.000	0.000	0.000	1506.232	0.000	0.000	1561.327
2034	20.995	5.731	0.000	26.726	1130.836	310.953	0.000	0.000	0.000	0.000	0.000	1441.789	0.000	0.000	1468.516
2035	15.742	5.209	0.000	20.951	1080.077	301.338	0.000	0.000	0.000	0.000	0.000	1381.414	0.000	0.000	1402.366
2036	8.503	4.695	0.000	13.199	1031.599	291.907	0.000	0.000	0.000	0.000	0.000	1323.507	0.000	0.000	1336.706
2037	6.749	4.231	0.000	10.980	983.968	281.785	0.000	0.000	0.000	0.000	0.000	1265.753	0.000	0.000	1276.733
2038	15.860	3.618	0.000	19.478	935.436	268.956	0.000	0.000	0.000	0.000	0.000	1204.392	0.000	0.000	1223.870
2039	5.379	3.505	0.000	8.884	881.700	250.828	0.000	0.000	0.000	0.000	0.000	1132.528	0.000	0.000	1141.412
2040	9.373	3.384	0.000	12.757	842.784	242.349	0.000	0.000	0.000	0.000	0.000	1085.133	0.000	0.000	1097.890
2041	-13.251	2.816	0.000	-10.435	802.638	234.549	0.000	0.000	0.000	0.000	0.000	1037.187	0.000	0.000	1026.752
TOTALS	1,159.205	155.679	0.000	1,314.885	26,346.280	6,904.528	0.000	0.000	0.000	0.000	0.000	33,250.809	0.000	2.901	34,568.595

Discounted costs of investment scenarios

The second table for each scenario calculates the carbon costs associated with the emissions calculated in the tool kit. The cost of carbon is defined in the ***Input Data*** sheet.

Discounted Emissions Costs (in millions of US Dollar)

Works	Vehicle running (ICE)	Vehicle running (EV)	Vehicle manufacture	Other (from External Calculations)	CO2 Reduction	Total CO2 Costs
70.923	208.153	0.170	83.036	0.004	-0.042	362.244
46.356	155.090	0.152	65.031	0.008	0.000	266.636
3.425	146.371	0.143	61.197	0.010	-0.035	211.112
9.491	139.737	0.137	58.769	0.013	0.000	208.147
2.281	133.010	0.131	55.949	0.014	-0.029	191.357
2.343	126.771	0.125	53.482	0.016	0.000	182.736
5.925	120.845	0.119	51.068	0.017	-0.024	177.951
3.847	115.309	0.114	48.702	0.017	0.000	167.990
2.620	110.029	0.109	46.426	0.000	-0.020	159.166
9.899	105.460	0.105	44.278	0.000	0.000	159.741
5.909	101.537	0.101	41.752	0.000	0.000	149.300
20.614	96.978	0.097	40.148	0.000	0.000	157.837
13.179	92.132	0.093	38.374	0.000	0.000	143.777
7.752	87.654	0.089	36.638	0.000	0.000	132.132
5.879	83.496	0.085	34.880	0.000	0.000	124.339
4.065	79.418	0.081	33.214	0.000	0.000	116.778
7.459	75.310	0.077	31.641	0.000	0.000	114.487
3.903	71.135	0.073	29.737	0.000	0.000	104.848
5.672	67.657	0.070	28.517	0.000	0.000	101.916
1.852	64.289	0.067	27.205	0.000	0.000	93.412
233.394	2,180.380	2.137	910.044	0.099	-0.149	3,325.906

Discounted emission costs

This sheet enables you to view the detailed economic data associated with each of the scenarios and analyze the annual cost streams. It also allows you to compare the carbon cost of each scenario and where the costs are being generated.

Impact of road user costs associated with transition to EVs can be included through HDM-4 analysis. This requires calibration of road user cost models and adaptation of vehicle fleet inputs in HDM-4 for EVs. An example of HDM-4 EV road user cost models' calibration factors is provided in Appendix 1.

D.10 Output: Total Emission Graphs

This sheet gives you the graphical and tabular reports of the total emissions calculated for the five scenarios. The results of each scenario are presented in a graph.

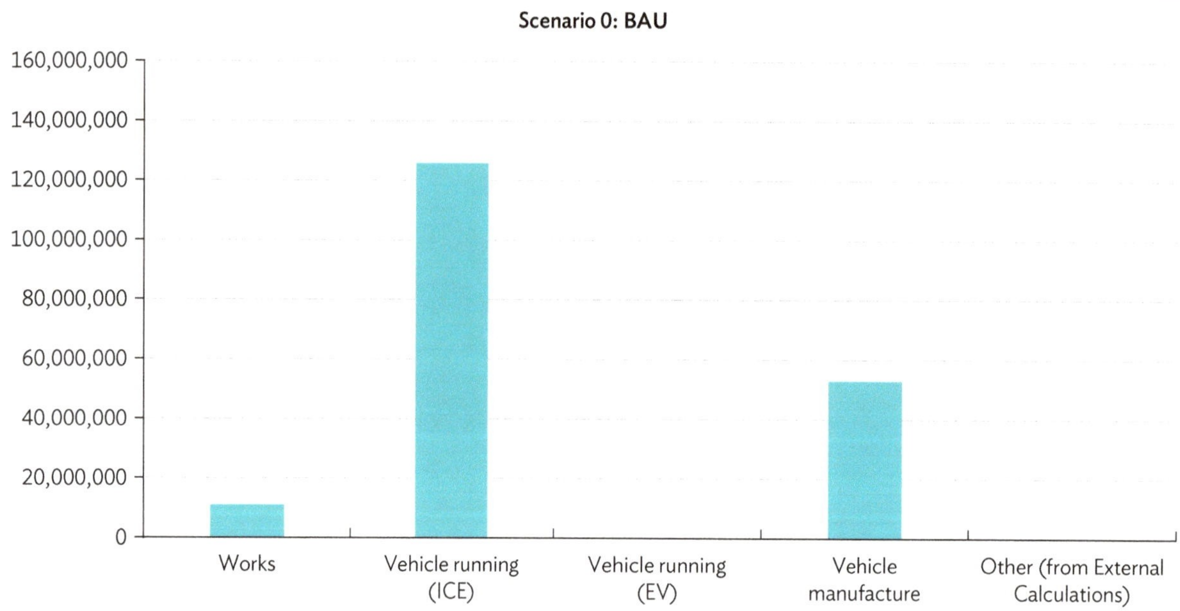

EV = electric vehicle, ICE = internal combustion engine.

Total emissions by scenario and category

This graph illustrates the magnitude of the emissions for each of the emission components in the tool kit.

The tabular data enables you to quickly compare the emissions produced by each of the five scenarios. The volume of the emissions is shown below, including percentage distribution.

Scenario Name	**Scenario 0: BAU**	**Scenario 1: WP+EV Charging**	**Scenario 2: WP+EV Charging+Bridge**	**Scenario 3: WP+EV Charging+Tunnel**	**Scenario 4: WP+EV Charging+Bridge**
Scenario Option	**Budget (100%)**	**Budget (80%)**	**Budget (60%)**	**Budget (40%)**	**Budget (100%)**
	Tonnes of CO2				
Works	11,218,826	9,784,568	9,725,550	8,183,267	11,218,826
Vehicle running (ICE)	125,944,618	124,441,156	131,857,321	137,829,938	125,944,618
Vehicle running (EV)	126,604	127,759	130,353	136,153	126,604
Vehicle manufacture	52,994,665	55,547,182	57,010,386	62,635,771	53,005,696
Other (from External Calculations)	3,600	7,327	7,200	7,200	39,727
TOTAL	**190,288,313**	**189,907,992**	**198,730,810**	**208,792,329**	**190,335,471**
Carbon Reduction Scheme	5,000	20,000	5,030	60	80
	Percentage (%)				
Works	5.90%	5.15%	4.89%	3.92%	5.89%
Vehicle running (ICE)	66.19%	65.53%	66.35%	66.01%	66.17%
Vehicle running (EV)	0.07%	0.07%	0.07%	0.07%	0.07%
Vehicle manufacture	27.85%	29.25%	28.69%	30.00%	27.85%
Other (from External Calculations)	0.00%	0.00%	0.00%	0.00%	0.02%

Tabular emission summary by scenario and category

This tabular data highlights that the road maintenance activities (the works) account for less than 6% of the total emissions produced. The vehicle running emissions account for most of the emissions produced at about 66%.

D.11 Output: Annual Emissions

Annual CO2 Emissions by Source

Option	Source	Total	% of Total	1 2022	2 2023	3 2024	4 2025	
Scenario 0: BAU	Works	11,218,826	6%	1,688,635	1,214,093	98,669	300,777	
	Vehicles running (ICE/HYBRID)	126,658,294	66%	4,956,025	4,061,872	4,216,886	4,428,331	4
	Vehicle running (EV)	126,604	0%	4,042	3,969	4,117	4,336	
	Vehicle manufacture	52,994,665	28%	1,977,056	1,703,192	1,763,050	1,862,428	1
	Other (from External Calculations	3,600	0%	100	200	300	400	
	Sub-Total	**191,001,989**	**100%**	**8,625,858**	**6,983,325**	**6,083,022**	**6,596,273**	**6**
	CO2 Reduction			1,000	0	1,000	0	
	TOTAL			**8,624,858**	**6,983,325**	**6,082,022**	**6,596,273**	**6**

Annual emissions by category

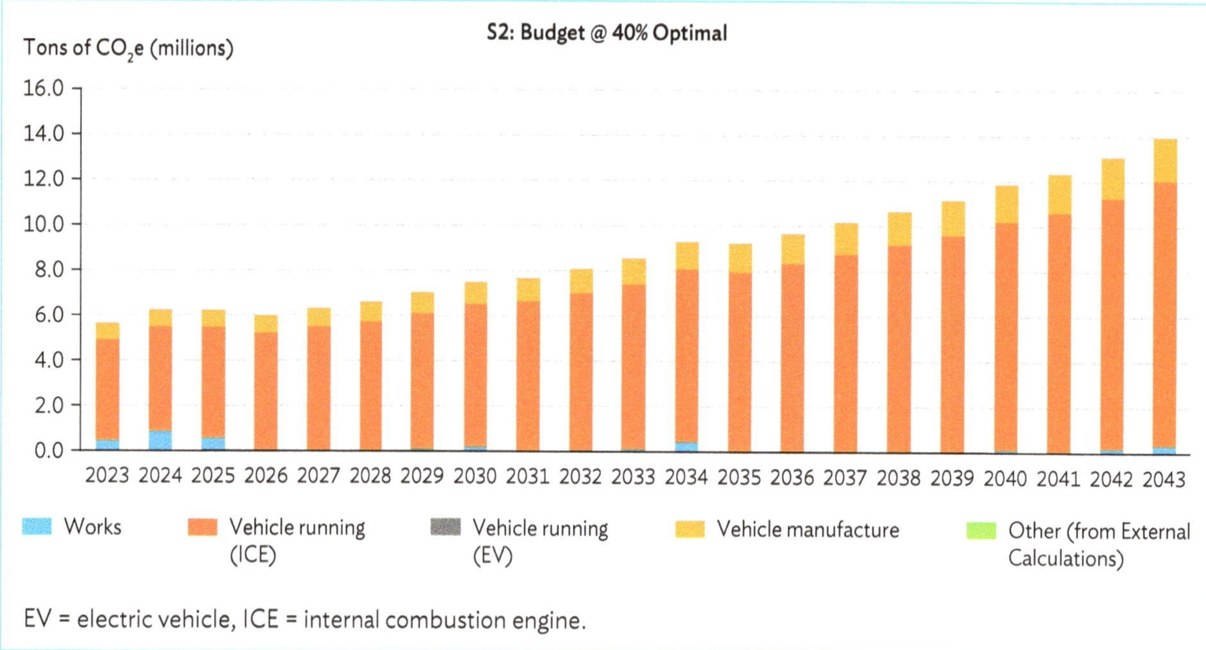

EV = electric vehicle, ICE = internal combustion engine.

Graph of annual emissions by category

D.12 Output: Annual Emission Graphs

This sheet gives you the graphical reports of the annual emissions calculated for one of the selected scenarios. The results of each emission category are presented in a separate graph:

(1) Annual emission savings from the carbon reduction scheme

(2) Annual emission from work activities

(3) Annual emissions from vehicle running

(4) Annual emissions from electric vehicles

(5) Annual emissions from vehicle manufacturing

(6) Annual total emissions

To select the scenario to display in the reports, highlight the ***Select Budget Scenario for Option*** input, click the down arrow, and click on the required scenario.

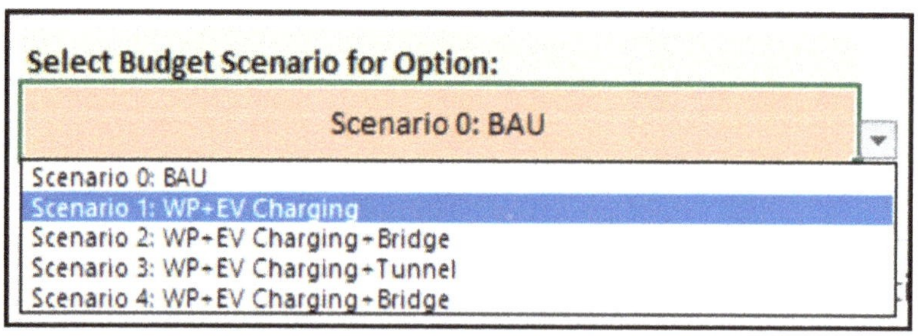

Selecting scenarios to display on graph

The graphs will automatically update the results for the selected scenario.

Annual emissions by category and scenario

This graph provides a snapshot of the results of a single scenario, allowing the analysis of the annual carbon trend and magnitude of individual contributions. The total carbon emissions for the scenario can be viewed alongside the carbon savings from the carbon reduction scheme implemented, to determine whether this is sufficient to meet the objectives of the road investment scheme under analysis.

D.13 Output: Annual Vehicle Data

This sheet gives you the tabular and graphical reports of the annual emissions calculated together with the average road condition and vehicle speed for each scenario. The annual emissions for each of the vehicles are also included.

Annual CO2 Emissions and Vehicle Speeds								
Scenario	Option	Source			Total	1 2022	2 2023	3 2024
Scenario 0: BAU	Budget (100%)	Avg Vehicle Speed (km/h)				50.91	57.08	64.54
		Avg. IRI (m/km)				11.42	8.04	5.65
		Vehicle Emissions (CO2 tonnes)	F- Medium Truck 3-A:	ICE	5,880,566	891,324	173,205	181,009
			E- Small 2-axle Truck	ICE	4,118,816	157,942	141,223	147,206
			H- Articulated Vehicl	ICE	19,232,850	707,934	628,484	659,413
			D- Van up to 2t	ICE	14,070,183	536,853	487,138	500,656
			C- Bus	ICE	30,601	1,119	1,039	1,087
			B- Minibus	ICE	7,962,063	299,128	276,812	284,983
			A- Personal Car	EV	128,484,484	4,342,823	4,061,131	4,209,700
		Vehicle Emissions (CO2 tonnes)			179,779,563	6,937,123	5,769,032	5,984,053
		Works, CO2 (tonnes)			11,218,826	1,688,635	1,214,093	98,669

Tabular annual emissions, vehicle speeds, and road condition

The graphs show the total emissions from vehicles and works, plotted against the average vehicle speed.

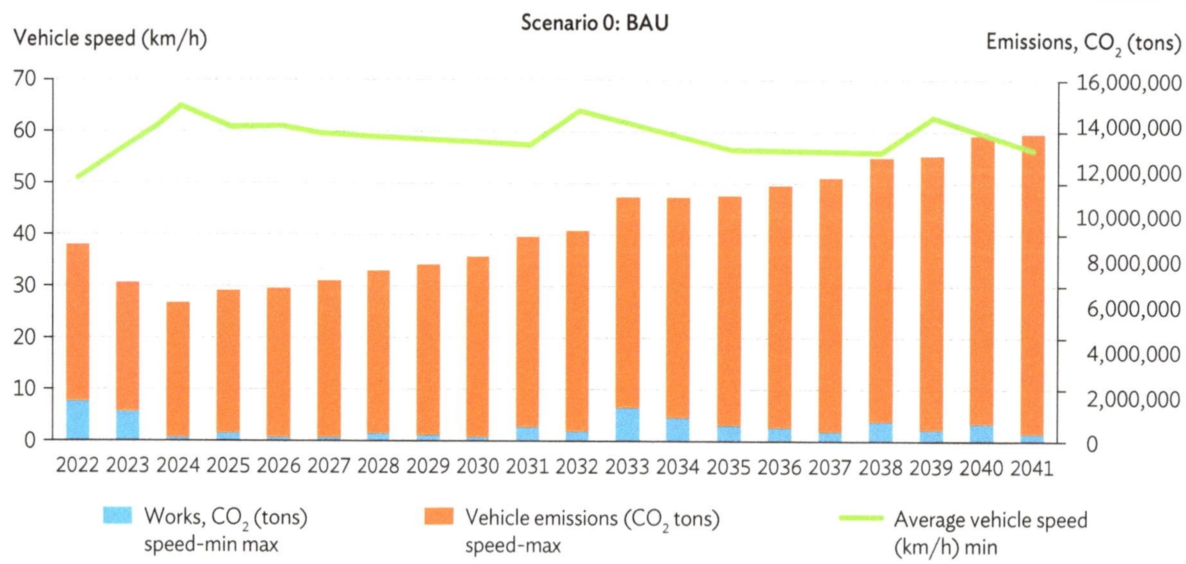

CO_2 = carbon dioxide, km/h = kilometer per hour.

Vehicle speed compared with annual works and vehicle emissions

This report is useful for understanding the relationship between vehicle speed, road network condition, and emissions. The impact of these can be seen on the works and total vehicle emissions, as well as on each individual vehicle type. This enables you to see which vehicle types are contributing the most to the total vehicle emissions, and how they are impacted by average vehicle speed and condition.

When comparing the annual trends of emissions, it should be noted that, often, as traffic volume continues to grow each year, emissions also tend to increase over time. The growth of traffic is defined as part of the HDM-4 analysis.

D.14 Output: Vehicle Speeds

This sheet gives you the tabular and graphical reports of the annual emissions calculated together with the average road condition and vehicle speed for each scenario.

This is a similar report to ***Output – Annual Vehicle Data*** but does not include detailed vehicle emissions. It is easier to see the data overview without the extra details that are not required.

Scenario	Source	1	2	3	4	5
		2022	2023	2024	2025	2026
Scenario 0: BAU	Avg Vehicle Speed (km/h	50.91	57.08	64.54	60.71	60.73
	Avg. IRI	11.42	8.04	5.65	6.61	6.58
	Vehicle Emissions, CO2	6,937,123	5,769,032	5,984,053	6,295,096	6,591,590
	Works, CO2 (tonnes)	1,688,635	1,214,093	98,669	300,777	79,519

Annual vehicle and works emission, and vehicle speed and condition

These graphs are like the output ***Annual Vehicle Data***, and are included to enable a high-level comparison.

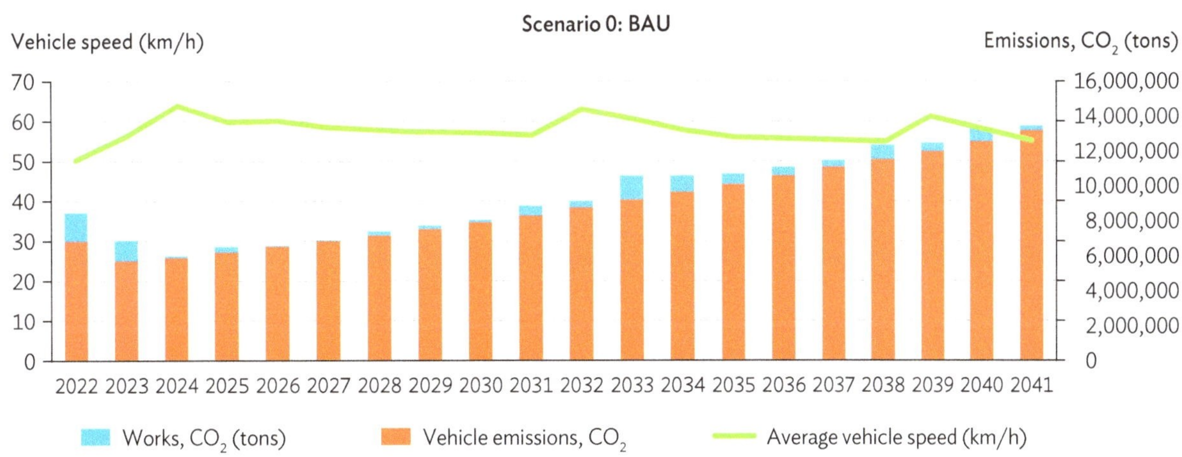

CO_2 = carbon dioxide, km/h = kilometer per hour.

Annual vehicle and works emission compared with average speed

D.15 Output: Economics

This sheet gives you the tabular and graphical report of the economic analysis.

Discounted Cost (in millions of US Dollar)

		Road Agency Cost (RAC)			Road User Costs (RUC)				Other			
Scenario	Budget/Alternative	Capital Costs	Recurrent Costs	Special Costs	VOC	TTC	Accident Costs	Emission Costs	Exogeneous Costs	Additional Costs	Total Transport Costs	Total Transport Costs (excluding CO_2 Costs)
S1: Business as Usual	5. BAU (Based on 2019 Budget)	335.607	267.406	0.000	40,060.388	11,324.574	0.000	3,706.535	0.000	0.000	55,694.511	51,987.976
S2: Budget @ 40% Optimal	4. Reduced Budget (less 60%)	808.608	289.243	0.000	38,365.931	10,044.367	0.000	3,753.364	0.000	0.000	53,261.513	49,508.149
S3: Budget @ 60% Optimal	3. Reduced Budget (less 40%)	1,256.247	305.552	0.000	36,870.178	9,090.478	0.000	3,717.118	0.000	0.000	51,239.574	47,522.455
S4: Budget @ 80% Optimal	2. Reduced Budget (less 20%)	1,636.396	285.849	0.000	36,171.258	8,697.093	0.000	3,724.283	0.000	0.000	50,514.880	46,790.597
S5: Budget @ Optimal	1. Optimal Budget	2,469.971	287.017	0.000	34,982.229	7,939.937	0.000	3,735.461	0.000	0.000	49,414.614	45,679.153

◀ **Summary of economic analysis – discounted costs (US$ million)**

For the sake of economic indicators, scenario 1 is considered the base case against which other scenarios are compared to.

The net present value (NPV) is given with and without the carbon cost included.

NPV	NPV (excluding CO_2 Costs)	Incremental BCR	Incremental BCR (excluding CO_2 Costs)
–	–	1	1
185.083	148.295	1.141	1.113
−1,123.308	−1,000.748	0.039	0.144
−3,864.217	−3,585.859	−3.476	−3.154
31.122	2.747	1.057	1.005

▲ **Net present value and benefit-cost ratios**

The graph shows the magnitude of the discounted economic costs for each of the five scenarios.

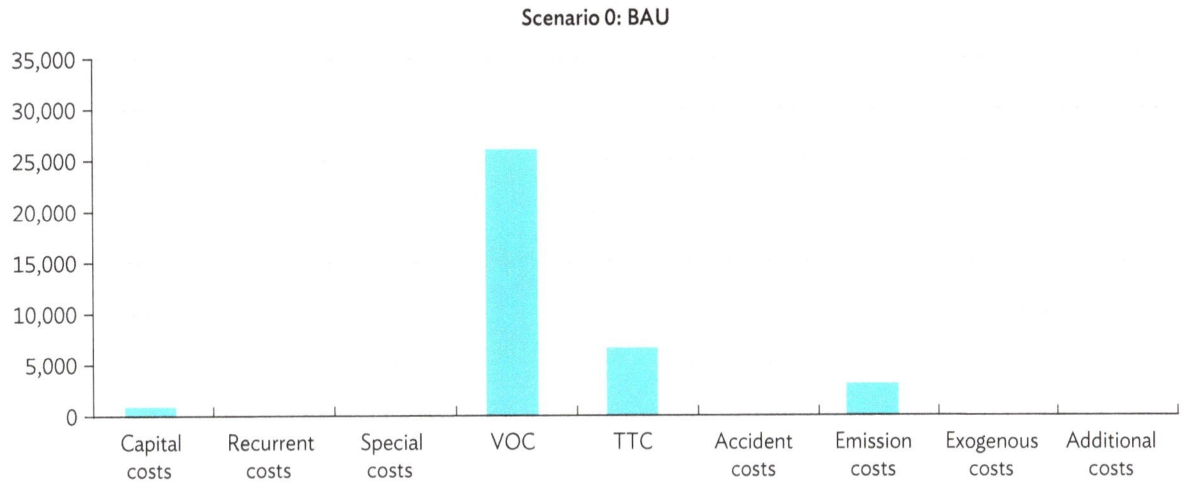

TTC = Travel Time Cost, VOC = Vehicle Operating Costs.

Discounted economic costs for each scenario

This report is useful in comparing the economic costs of the five scenarios and the impact on the economic indicators of including the carbon cost for the emissions evaluated in the tool kit.

This report can be used alongside the reports of the carbon emissions to determine the scenario that best meets the objective. The objective may vary from scheme to scheme; however, typically, the aim is to minimize the carbon impact while maximizing the economic benefit of the road investment.

E. ILLUSTRATIVE APPLICATION OF THE TOOL KIT

E.1 Introduction

This section illustrates a network analysis performed using Strategy Analysis in HDM-4 version 2. The data in the example given is for illustrative purposes only.

E.2 HDM-4 Setup

To perform this analysis, Strategy Analysis will be used. The aim of this study is to determine the emissions impact of network maintenance strategy under different budget constraints.

Prior to completing the HDM-4 analysis, the user will gather the data required to set up HDM-4 correctly. The data requirements are beyond the scope of this manual; therefore, it is assumed that the user has some familiarity with HDM-4 and the concepts.

E.2.1 Vehicle Fleet

The vehicle fleet used for this analysis consists of seven vehicles to represent the typical vehicles using the road network under analysis. The vehicle types include cars, buses, vans, and lorries. In this analysis, they are all ICE vehicles. It is assumed that the calibration of these vehicles is suitable for the network being analyzed. There are no nonmotorized vehicles in the vehicle fleet.

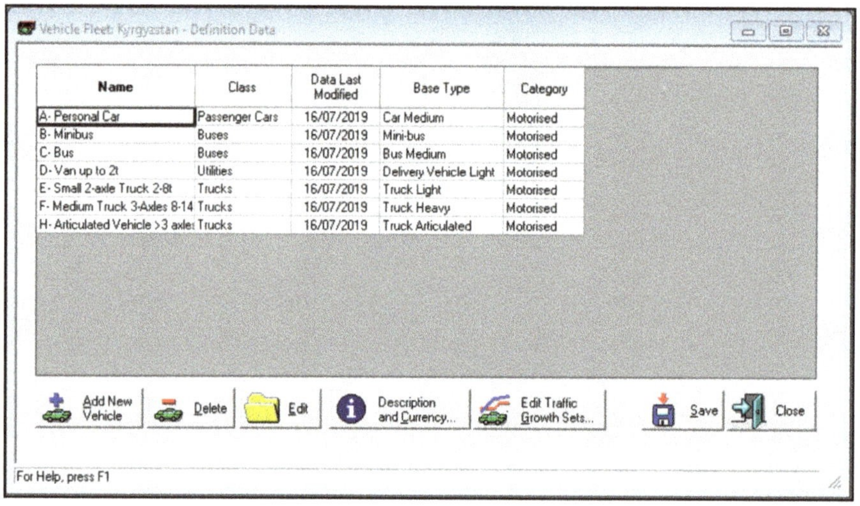

E.2.2 Road Network

The road network being analyzed is defined in a road network. As this is a network-level analysis, representative sections are used to model the total network. The network includes unsealed and bituminous road sections.

For each of the representative sections, the traffic volume and condition are defined, as well as the characteristics of the section (length, geometry, strength, pavement thicknesses, etc).

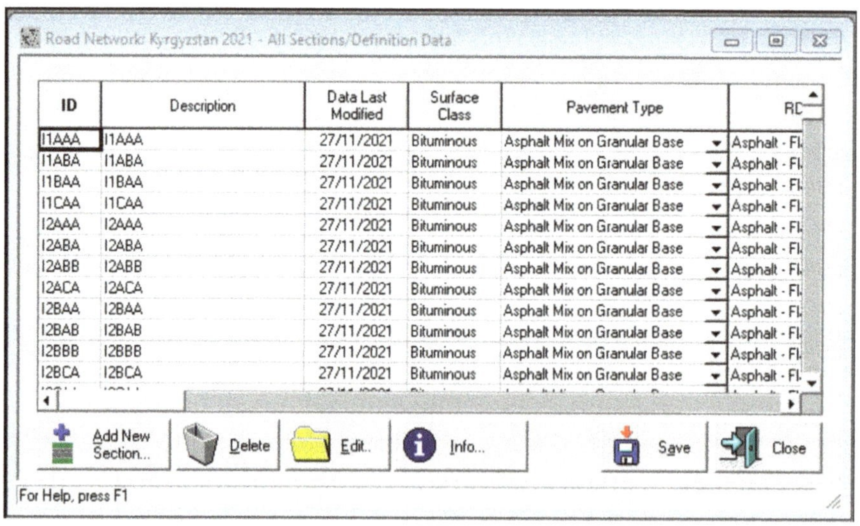

E.2.3 Strategy Analysis

To perform the network-level analysis, a scenario analysis is created and connected to the road network and vehicle fleet previously defined. The start year is set to 2022, and the duration of the analysis is set to 20 years.

The currencies used in the analysis are all the same, so the conversion factors are all set to one. The discount rate for the economic evaluation is set at 10%.

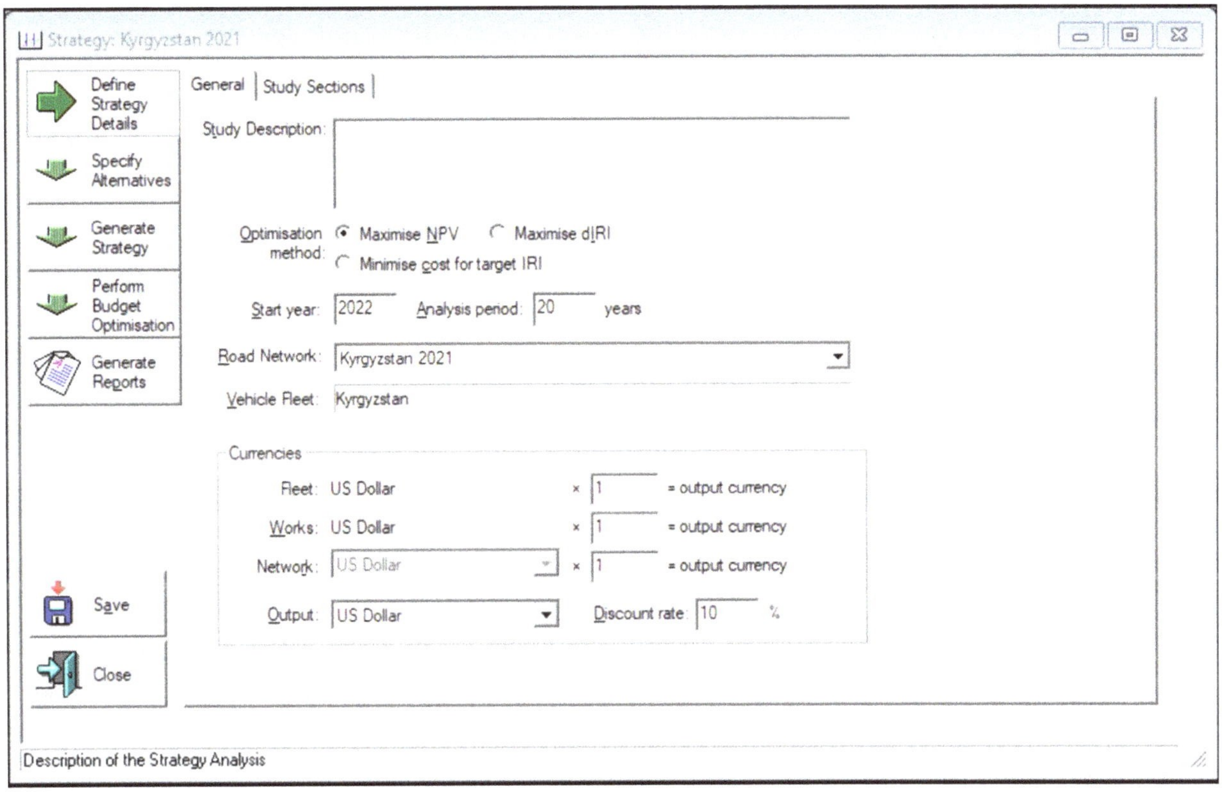

The required sections within the road network are selected for inclusion in the analysis. The total number of sections is 223. It should be noted that the tool kit does not report at the section level or section-aggregate level, just at the network level.

Each of the selected sections is assigned a traffic growth set that tells HDM-4 how the traffic volumes will change over time.

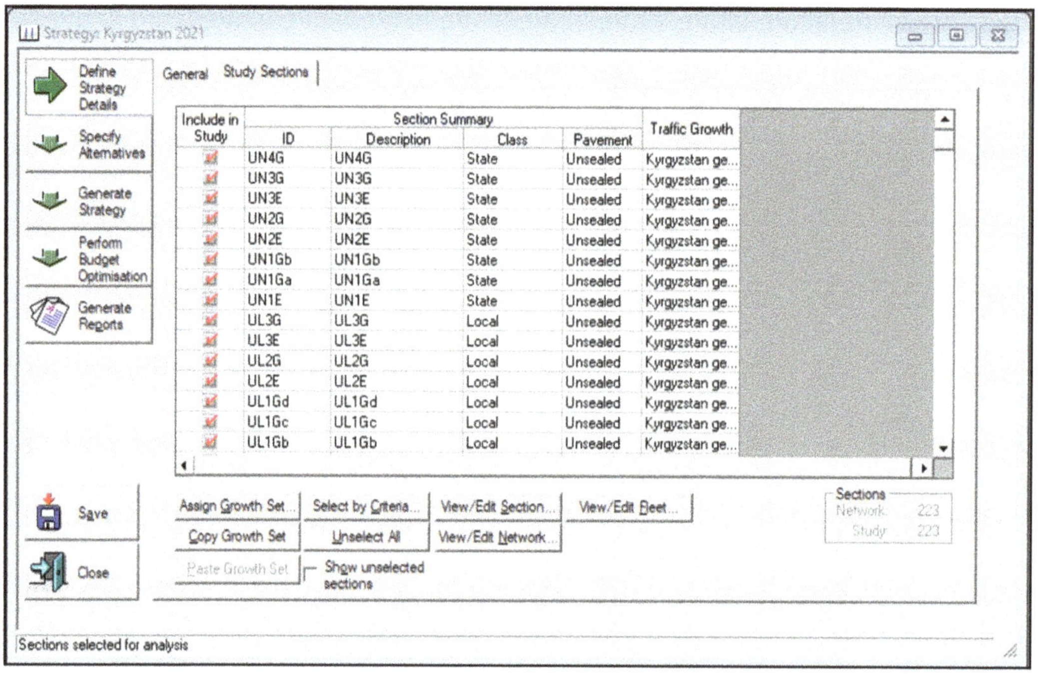

After selecting the sections, set up the alternative ways in which you wish HDM-4 to consider maintaining and improving each of the representative sections. In this example, the alternatives were set up to perform routine maintenance every year, but with the option to perform rehabilitation activities (overlays) in multiple years. In this way, when budget optimization is performed, HDM-4 has multiple options for selecting an alternative set of road maintenance interventions to apply in a year wherein the budget is available for spending.

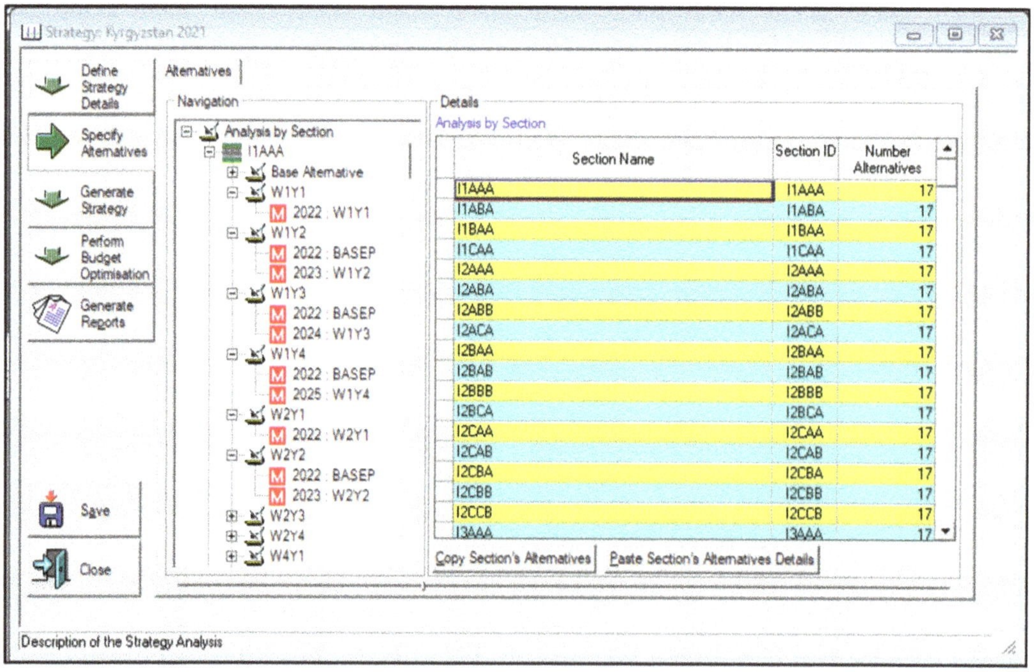

Before running the initial analysis, the run must be set up with the correct options to allow the tool kit to read the required data. The important model selections are as follows:

- Energy Balance
- Emissions
- Include annual vehicle data (therefore the Exclude box must be unticked as shown below)

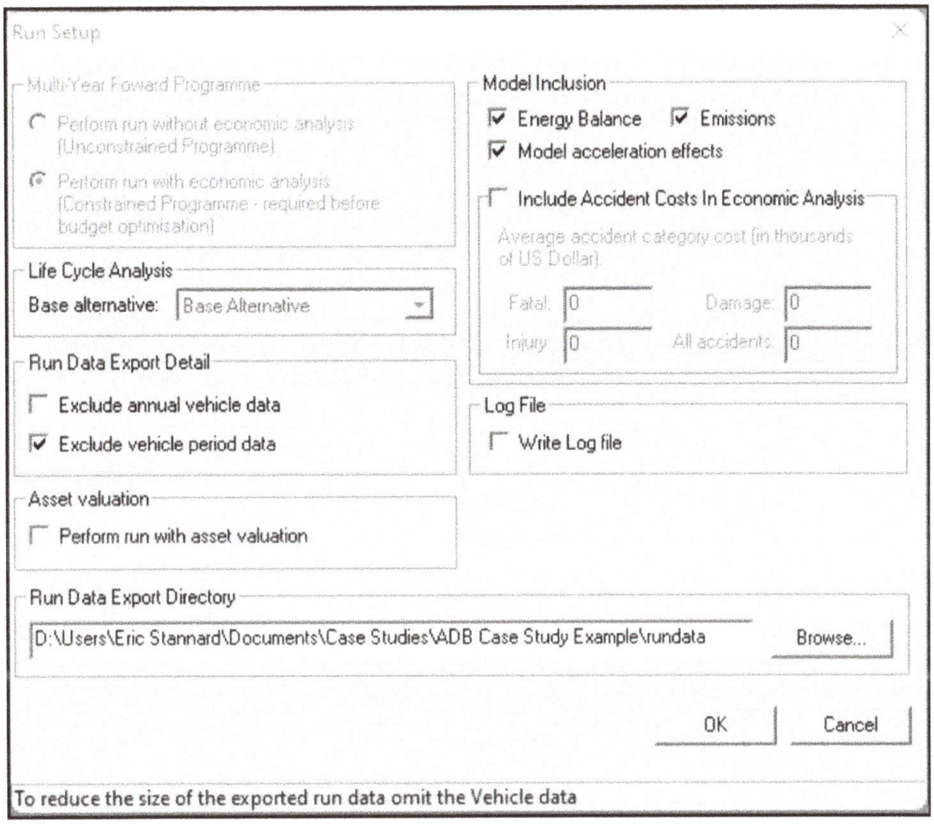

You should make a note of the run data export directory defined as this is required when the tool kit is run.

When the analysis is run successfully, the user can move on to define the budget scenarios they wish to include the results for. The tool kit can only show five scenarios at a time, but you can define more and change the selections in the tool kit for reporting.

In this example, different budget scenarios have been defined based on an unconstrained budget:

- Budget 100%
- Budget 80%
- Budget 60%
- Budget 40%

The budget scenario optimization should be performed successfully before moving on to using the tool kit.

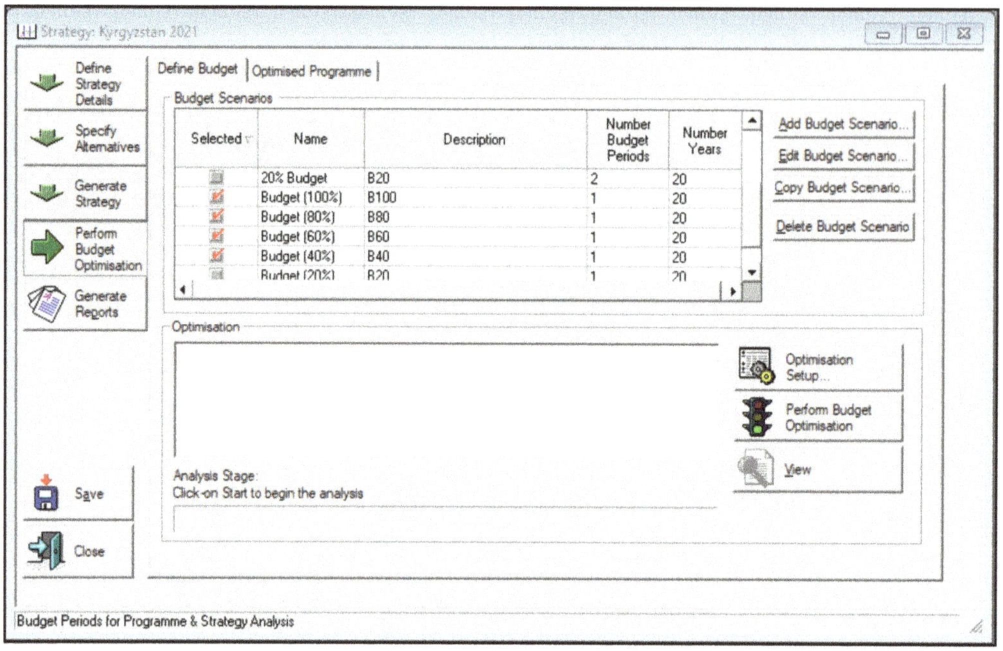

Prior to using the tool kit, the user should check the results in HDM-4 to ensure that they are correct.

E.3 Importing Data to the Tool Kit

Once the analysis is complete in HDM-4, the tool kit can be opened.

To import the data into the tool kit, carry out the following steps:

- Go to the ***Input Tables*** sheet and enter the location of the HDM-4 run data into the box (or use the ***Browse*** button to locate it)

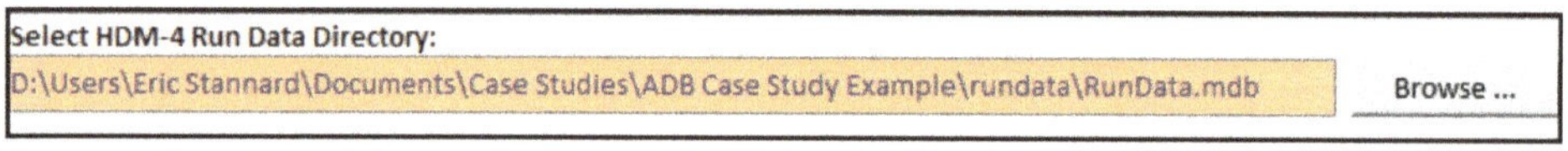

- Click on the ***Step 2*** box. The HDM-4 data will be imported into the tool kit in the required format. The project title will be updated, and an overview of the analysis will be displayed:

Run Data	
Number of Budget Scenarios	4
Number of Years	20
Road length (km)	18,686

- Click on the ***Step 3*** box. The tool kit will be initialized with the vehicle fleet and works information; the sheets will be configured to show the correct number of years, works, and vehicles.

The data import and configuration of the tool kit is now complete and ready for the input tables to be updated before the analysis can be reviewed.

Each time the data is changed in the HDM-4 analysis and a new run is performed, this initial stage must be performed to ensure the tool kit is using the latest data.

In this example, the network data has been imported for an analysis of 18,686 km of roads over an analysis period of 20 years. The user has specified four budget scenarios.

E.4 Setting Up the Data

E.4.1 Input Tables

There are three key input tables you must enter data for:

- Emissions from Works
- Vehicle Manufacture and Fuel
- Economic Parameters

Emissions from Works

Number of Work Items:		26	
Maximum Number of Work Items Allowed		100	

WORK_CODE	WORK_NAME	kg of CO2 /unit	Unit
	Prep. Edge Repair	35	sq. m
	Prep. Patching	50	sq. m
	Prep. Spot Regravelling	10	cu. m
CS	Crack Sealing	30	sq. m
G2Y1	G2Y4	324000	km
G2Y2	G2Y1	324000	km
G2Y3	G2Y2	324000	km
G2Y4	G2Y3	324000	km
grad	grading	1000	km
GRAV	Regravelling	45	cu. m
OL	Overlay	45	sq. m
OL	Overlay 4cm	45	sq. m
OL	Overlay 8 cm	45	sq. m
PM	Preventive Overlay	45	sq. m
PO	Preventive Overlay	45	sq. m
POPA	Patching	45	sq. m
R1	Reconstruction C1	45	sq. m
R2	Reconstruction C2	45	sq. m
R3	Reconstruction C3	45	sq. m
Rec	Reconstruction	45	sq. m
REGR	Regravelling2	45	cu. m
REGR2	Gravelbase	45	cu. m
RM	Routine Maintenance	300	km
SDSC	Surface Dressing	45	sq. m
spot2	spot regravel2	20	cu. m
SPOT2	spot2	20	cu. m

Note that when entering the kilogram of CO_2 per unit, the units vary between square meter, cubic meter, and kilometer. It is important to enter the correct value for the correct units.

Vehicle Manufacture & Fuel

Number of Vehicles: 7
Well-Tank Adjustment Factor: 1.3

Vehicle ID	Vehicle Name	kg of CO2 /vehicle	ICE/EV/NMT	HDM-4 CO2 Included	Vehicle Category
1	F- Medium Truck 3-A	70,000	ICE	100%	Medium Truk
2	E- Small 2-axle Truck	60,000	ICE	100%	Light Truck
3	H- Articulated Vehic	100,000	ICE	100%	Heavy Truck
4	D- Van up to 2t	35,000	ICE	100%	SUV/Utility
5	C- Bus	70,000	ICE	100%	Bus
6	B- Minibus	50,000	ICE	100%	SUV/Utility
7	A- Personal Car	25,000	EV	100%	Car

Economic Parameters

Discount Rate	10%
Currency	US Dollar
CO2 Costs/tonne	42
Currency	US Dollar

E.4.2 Scenarios

The final step to set up the tool kit before looking at the results is to define the scenarios. For the simplest use of the tool kit, the following need to be selected for each of the scenarios:

- Scenario Name
- HDM-4 Investment Alternative

If you are using electric vehicles, then you must also enter:

- Grid Factor

For this analysis, there are four HDM-4 investment alternatives, so the scenarios are set up as follows:

No.	Scenario Name	HDM-4 Investment Alternative
1	Business as Usual	Budget 100%
2	Reduced Budget 80%	Budget 80%
3	Reduced Budget 60%	Budget 60%
4	Reduced Budget 40%	Budget 40%

E.5 Reviewing the Results

Looking at the total emissions (***Output-TotalEmissionGraphs***) for the scenarios, we can see that as the budget is restricted, the emission from maintenance works decreases (because less work is being carried out) but the emission from vehicles increases.

The total emissions (Works + Vehicles) show that the funding between 100% and 80% of the total network budget required for maintenance produces similar total emissions. However, as the budget is further reduced, the total emissions begin to climb.

It can also be seen that for this network, the maintenance activities account for between 3% and 6% of the total emissions associated with the network. Therefore, the significant proportion of the emissions is generated by the vehicles using the network, 94% to 97%.

Scenario Name	**Business As Usual**	**Reduced Budget 80%**	**Reduced Budget 60%**	**Reduced Budget 40%**
Scenario Option	**Budget (100%)**	**Budget (80%)**	**Budget (60%)**	**Budget (40%)**
			Tonnes of CO2	
Works	11,180,552	9,330,739	8,509,894	6,429,150
Vehicle running (ICE)	125,944,618	124,441,156	131,857,321	137,829,938
Vehicle running (EV)	0	0	0	0
Vehicle manufacture	52,994,665	55,547,182	57,010,386	62,635,771
Other (from External Calculations)	0	0	0	0
TOTAL	**190,119,835**	**189,319,078**	**197,377,601**	**206,894,859**
Carbon Reduction Scheme	0	0	0	0
			Percentage (%)	
Works	5.88%	4.93%	4.31%	3.11%
Vehicle running (ICE)	66.24%	65.73%	66.80%	66.62%
Vehicle running (EV)	0.00%	0.00%	0.00%	0.00%
Vehicle manufacture	27.87%	29.34%	28.88%	30.27%
Other (from External Calculations)	0.00%	0.00%	0.00%	0.00%

By looking at the annual reports of emissions, we can view the annual emission trend. It should be noted that during the life cycle of the analysis, the traffic volume is also growing, hence emissions are partly rising because of this traffic growth.

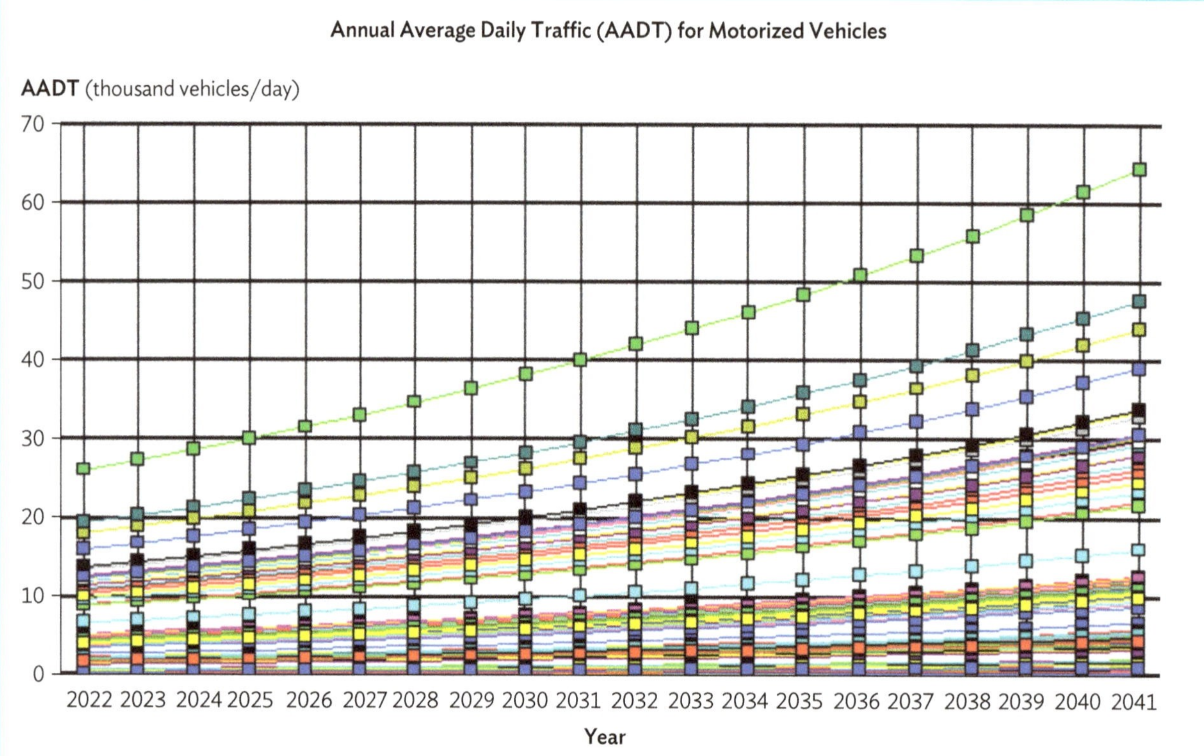

However, we can see that, initially, the emissions are dropping in the first few years of the analysis before starting to rise again.

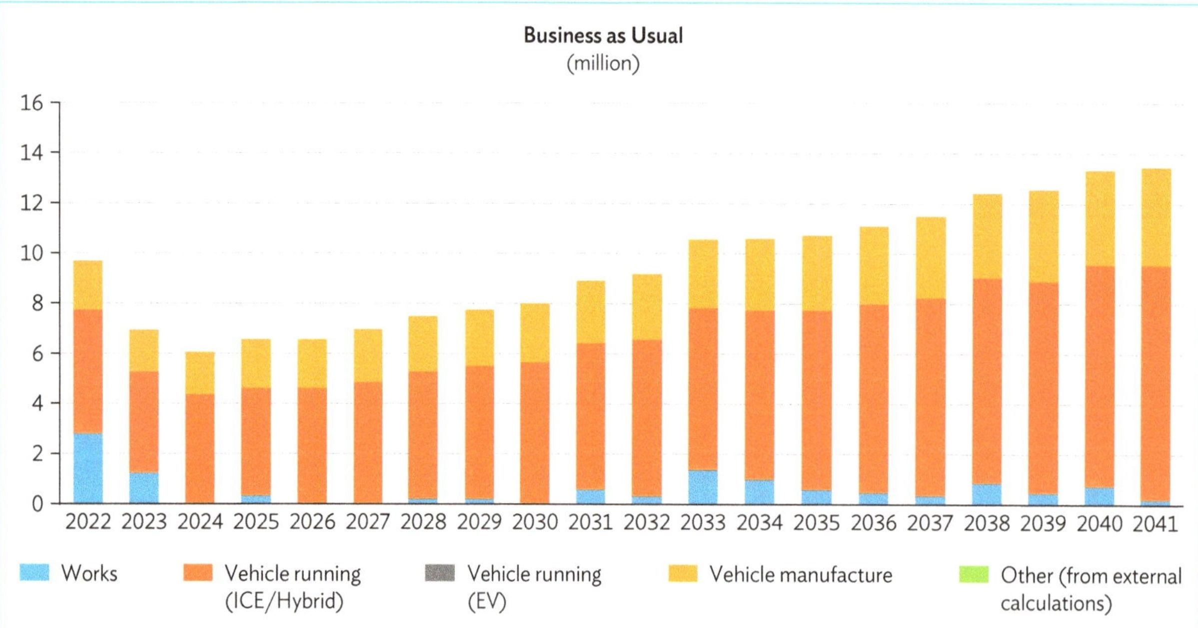

EV = electric vehicle, ICE = internal combustion engine.

We can see from the sheet, ***Analysis-HDM4WorksEmissions***, that, in the first few years of the analysis, there are a lot of reconstruction and overlay operations taking place, which will have a significant effect on network condition and network average speeds, which in turn will have an impact on the vehicle emissions.

Option	Source	Total	% of Total	1 2022	2 2023	3 2024	4 2025
Business As Usual	Works	11,180,552	6%	2,782,344	1,214,093	83,120	235,201
	Vehicles running (ICE/HYBRID)	126,658,294	66%	4,956,025	4,061,872	4,216,886	4,428,331
	Vehicle running (EV)	0	0%	0	0	0	0
	Vehicle manufacture	52,994,665	28%	1,977,056	1,703,192	1,763,050	1,862,428
	Other (from External Calculations	0	0%	0	0	0	0
	Sub-Total	**190,833,511**	**100%**	**9,715,425**	**6,979,157**	**6,063,057**	**6,525,960**
	CO2 Reduction			0	0	0	0
	TOTAL			**9,715,425**	**6,979,157**	**6,063,057**	**6,525,960**

This is further illustrated when we look at the emissions alongside the average vehicle speed, road network condition, and individual vehicle emissions (***Output-Annual Vehicle Data***).

Option	Source			Total	1 2022	2 2023	3 2024	4 2025
Budget (100%)	Avg Vehicle Speed (km/h)				50.91	57.08	64.54	60.71
	Avg. IRI (m/km)				11.42	8.04	5.65	6.61
	Vehicle Emissions (CO2 tonnes)	F- Medium Truck 3-A:	ICE	5,880,566	891,324	173,205	181,009	189,018
		E- Small 2-axle Truck	ICE	4,118,816	157,942	141,223	147,206	153,345
		H- Articulated Vehicl	ICE	19,232,850	707,934	628,484	659,413	690,755
		D- Van up to 2t	ICE	14,070,183	536,853	487,138	500,656	522,837
		C- Bus	ICE	30,601	1,119	1,039	1,087	1,134
		B- Minibus	ICE	7,962,063	299,128	276,812	284,983	297,343
		A- Personal Car	ICE	128,357,880	4,338,781	4,057,162	4,205,583	4,436,327
	Vehicle Emissions (CO2 tonnes)			179,652,959	6,933,080	5,765,064	5,979,936	6,290,759
	Works, CO2 (tonnes)			11,180,552	2,782,344	1,214,093	83,120	235,201

This report shows that as maintenance activities are carried out, average road condition improves, and the average speed of vehicles using the road increases. This has an impact on total vehicle emissions, causing a reduction from 6,933,080 CO_2 tons in the first year of analysis to 5,765,064 CO_2 tons in the second year. We can also look at the impact on individual vehicles.

The distribution of emissions shows that the largest percentage of emissions are produced by a personal car. However, the largest reduction of emissions following the initial maintenance works is the medium truck of which emissions go from 891,324 tons to 173,205 tons.

If we look at the economics, we can see the discounted carbon cost for the network during the analysis period. Most of the carbon costs are associated with vehicle use.

Works	Vehicle running (ICE)	Vehicle running (EV)	Vehicle manufacture	Other (from External Calculations)	CO2 Reduction	Total CO2 Costs
116.858	208.153	0.000	83.036	0.000	0.000	408.048
46.356	155.090	0.000	65.031	0.000	0.000	266.477
2.885	146.371	0.000	61.197	0.000	0.000	210.453
7.422	139.737	0.000	58.769	0.000	0.000	205.928
0.400	133.010	0.000	55.949	0.000	0.000	189.359
0.630	126.771	0.000	53.482	0.000	0.000	180.883
4.368	120.845	0.000	51.068	0.000	0.000	176.281
2.431	115.309	0.000	48.702	0.000	0.000	166.443
1.333	110.029	0.000	46.426	0.000	0.000	157.789
8.729	105.460	0.000	44.278	0.000	0.000	158.466
4.845	101.537	0.000	41.752	0.000	0.000	148.135
19.647	96.978	0.000	40.148	0.000	0.000	156.773
12.300	92.132	0.000	38.374	0.000	0.000	142.806
6.953	87.654	0.000	36.638	0.000	0.000	131.245
5.152	83.496	0.000	34.880	0.000	0.000	123.528
3.404	79.418	0.000	33.214	0.000	0.000	116.036
6.858	75.310	0.000	31.641	0.000	0.000	113.810
3.357	71.135	0.000	29.737	0.000	0.000	104.229
5.176	67.657	0.000	28.517	0.000	0.000	101.350
1.401	64.289	0.000	27.205	0.000	0.000	92.895
260.508	2,180.380	0.000	910.044	0.000	0.000	3,350.933

APPENDIX

HDM-4 ROAD USER COST CALIBRATION FACTORS FOR ELECTRIC CARS—A DEMONSTRATION

Description	Model Parameters	EC Value	ICEC Value
Fuel Consumption Model			
Idle engine speed: This is the minimum engine speed of the vehicle.	RPM-IDLE (rpm)	100	800
Engine speed parameters	RPM-a0 to RPM-a3	Default	Default
Fuel consumption when the vehicle is stopping	IDLE fuel consumption (mL/s)	0.063	0.36
Instantaneous fuel consumption to total power required by engine ZETAB: base fuel-to-power efficiency factor (mL/kW/s)	ZETAB (fuel to power efficiency factor) (mL/kW/s)	0.01675	0.067
Drive train efficiency	EDT	0.88	0.9
Decrease in engine efficiency factor	EHP	0.886	0.25
Percentage of total power of engine and accessories	PCTPENG	20%	80%
Rolling resistance coefficients (Kcr2) and engine and accessories power loss coefficients (Kpea)	Kcr2 and Kpea	1	1
Ratio of drag to power at 100 km/h: the proportion of the rated engine power utilized to overcome engine drag and accessories loading at 100 km/h	PACCS_a0	0.19	0.2
Oil loss due to contamination per 1,000 vehicle-km	Contamination losses (L/1,000 km)	0	0.4
Oil loss due to operation per 1,000 vehicle-km	Operation losses (L/1,000 km)	0	0.0028
Maintenance Model			
Calibration factor	Parts consumption rotational factor	0.5	1
Calibration factor	Labor hours rotational factor	0.5	1
Emissions Model			
Engine output and catalytic convertor	Tailpipe emissions	0	Default

continued on next page

Continued

Description	Model Parameters	EC Value	ICEC Value
Forces Model			
Rated engine power	Engine power (kW)	110	70
Maximum used driving power	Driving power (kW)	96.8	33
Maximum used braking power	Braking power (kW)	20	20
Basic Characteristics			
Average number of kilometers driven per year	Utilization (km)	16,100	23,000
Hours per year expended on the essential tasks of complete round trips	Working hours (hrs)	402	550
Average vehicle service life	Average life (yrs)	10	10
Average operating weight of the vehicle type	Operating weight (tons)	1.6	1.2
Economic Unit Costs			
Average purchase cost of a new vehicle of this type	New vehicle price ($)	37,418	28,000
Average fuel cost per liter	Fuel price ($)	1.77	1.83
Average cost of maintenance labor	Maintenance labor ($)	23.8	23.8
Average cost of lubricants per liter	Lubricating oil ($)	3.45	3.45

h = hour, hrs = hours, km = kilometer, kW = kilowatt, mL = milliliter, yrs = years.

Source: Abdalla, A. 2021. Electric Vehicles Road User Costs Using HDM-4. University of Birmingham, United Kingdom.

REFERENCES

Asian Development Bank (ADB). 2010. *Methodology for Estimating Carbon Footprint of Road Projects*.

ADB. 2017. Guidelines for the Economic Analysis of Projects.

Berners-Lee, M. 2010. *How Bad are Bananas? The Carbon Footprint of Everything*.

Berners-Lee, M., and D. Clark. 2010. What's the carbon footprint of ... a new car? *The Guardian*. 23 September.

Brennan, J. W., and T. E. Barder. 2016. *Battery Electric Vehicles vs. Internal Combustion Lifecycle CO_2 Emissions*.

Buberger, J. et al. 2022. *Total CO_2-equivalent life-cycle emissions from commercially available passenger cars*.

Carbon Brief. 2020. *Factcheck: How electric vehicles help to tackle climate change*.

Crippa, M. et al. 2020. Fossil CO_2 Emissions of All World Countries. *JRC Science for Policy Report*. Publications Office of the European Union, Luxembourg.

Grutter, J., and K.-J. Kim. 2019. E-Mobility Options for ADB Developing Member Countries. *ADB Sustainable Development Working Paper Series*. No. 60.

Henning, T. F. P. 2023. Unpublished.

Hoekstra, A. 2020. *Producing gasoline and diesel emits more CO_2 than we thought*.

Intergovernmental Panel on Climate Change (IPCC). 2021. Climate Change 2021: The Physical Science Basis. Contribution of Working Group I to the Sixth Assessment Report of the Intergovernmental Panel on Climate Change. Chapter 7, Section 7.6.1.1. Cambridge: Cambridge University Press, United Kingdom.

Low Carbon Vehicle Partnership. 2018. *Lifecycle Emissions from Cars, Annex 1: Low CVP*. Press release.

National Treasury South Africa. 2023. Cities Infrastructure Delivery and Management System (CIDMS) Toolkit Addition 2. Annexure D Climate Resilience. National Treasury of South Africa. Pretoria, South Africa.

Nealer, R. 2015. *Cleaner Cars from Cradle to Grave: How Electric Cars Beat Gasoline Cars on Lifetime Global Warming Emissions*. Union of Concerned Scientists.

Patterson, J. et al. 2011. *Preparing for a Lifecycle CO_2 Measure*. Low Carbon Vehicle Partnership.

PE International. 2013. *Life Cycle CO_2e Assessment of Low Carbon Cars*. 2020–2030. Low Carbon Vehicle Partnership.

Qiao et al. 2017. *Electric vehicle recycling in China: Economic and environmental benefits*.

SLOCAT. 2021. *Tracking Trends in a Time of Change: The Need for Radical Action Towards Sustainable Transport Decarbonisation, Transport and Climate Change*. Global Status Report, 2nd edition.

Transport and Environment. 2022. *Update: T&E's Analysis of Electric Car*.

World Bank and ASTAE. 2010. *Greenhouse Gas Emissions Mitigation in Road Construction and Rehabilitation: A Toolkit for Developing Countries*. ROADEO Toolkit User Manual.

www.ingramcontent.com/pod-product-compliance
Lightning Source LLC
Chambersburg PA
CBHW041757220926
51847CB00014B/236
* 9 7 8 9 2 9 2 7 7 4 2 0 2 *